GIS Diffusion

The Adoption and Use of Geographical Information Systems in Local Government in Europe

Also in the GISDATA Series

Series Editors

I. Masser and F. Salgé

GIS Diffusion

The Adoption and Use of Geographical Information Systems in Local Government in Europe

EDITORS

IAN MASSER, HEATHER CAMPBELL,

and MASSIMO CRAGLIA

GISDATA III

SERIES EDITORS

IAN MASSER and FRANÇOIS SALGÉ

Taylor & Francis
Publishers since 1798

UK   Taylor & Francis Ltd, 1 Gunpowder Square, London EC4A 3DE
USA   Taylor & Francis Inc., 1900 Frost Road, Suite 101, Bristol, PA 19007

Copyright © Taylor & Francis Ltd 1996

British Library Cataloguing in Publication Data
A catalogue record for this book is available from the British Library.
ISBN 0-7484-0494-5 (cloth)
ISBN 0-7484-0495-3 (paperback)

Library of Congress Cataloging in Publication Data are available

Cover design by Hybert Design & Type
Typeset in Times 10/12pt by Keyword Publishing Services Ltd
Printed in Great Britain by T. J. Press (Padstow) Ltd

Contents

The GISDATA Series

Editors' Preface

Over the last few years there have been many signs that a European GIS community is coming into existence. This is particularly evident in the launch of the first of the European GIS (EGIS) conferences in Amsterdam in April 1990, the publication of the first issue of a GIS journal devoted to European issues (*GIS Europe*) in February 1992, the creation of a multi-purpose European ground related information network (MEGRIN) in June 1993 and the establishment of a European organization for geographic information (EUROGI) in October 1993. Set in the context of increasing pressures towards greater European integration, these developments can be seen as a clear indication of the need to exploit the potential of a technology that transcends national boundaries to deal with a wide range of social and environmental problems that are also increasingly seen as transcending national boundaries within Europe.

The GISDATA scientific programme is very much part of such developments. Its origins go back to January 1991 when the European Science Foundation funded a small workshop at Davos in Switzerland to explore the need for a European level GIS research programme. Given the tendencies noted above it is not surprising that participants of this workshop felt very strongly that a programme of this kind was urgently needed to overcome the fragmentation of existing research efforts within Europe. They also argued that such a programme should concentrate on fundamental research and it should have a strong technology transfer component to facilitate the exchange of ideas and experience at a crucial stage in the development of an important new research field. Following this meeting a small coordinating group was set up to prepare more detailed proposals for a GIS scientific programme during 1992. A central element of these proposals was a research agenda of priority issues grouped together under the headings of geographic databases, geographic data integration and social and environmental applications.

The GISDATA scientific programme was launched in January 1993. It is a 4-year scientific programme of the Standing Committee of Social Sciences of the European Science Foundation. By the end of the programme more than 300 scientists from 20 European countries will have directly participated in GISDATA activities and many others will have utilized the networks built up as a result of them. Its objectives are:

- To enhance existing national research efforts and promote collaborative ventures which aim to overcome European wide limitations in geographic data integration, databases design and social and environmental applications.

- To increase awareness of the political, cultural, organizational, technical and informational barriers to the increased utilization and inter-operability of GIS in Europe.

- To promote the ethical use of integrated information systems, including GIS, which handle socio-economic data by respecting the legal restrictions on data privacy at the national and European levels.

- To facilitate the development of appropriate methodologies for GIS research at the European level.

- To produce outputs of high scientific value.

- To build up a European network of researchers with particular emphasis on young researchers in the GIS field.

A key feature of the GISDATA programme is the series of specialist meetings that is being organized to discuss each of the issues outlined in the research agenda. The organization of each of these meetings is in the hands of a small task force of leading European experts in the field. The aim of these meetings is to stimulate research networking at the European level on the issues involved and also to produce high quality outputs in the form of books, special issues of major journals and other materials.

   With these considerations in mind the GISDATA series has been established in collaboration with Taylor & Francis Ltd to provide a showcase for this work. It has and will present the products of selected specialist meetings in the form of edited volumes of specially commissioned studies. The basic objective of the GISDATA series is to make the findings of these meetings accessible to as wide an audience as possible to facilitate the development of the GIS field as a whole. The books in the series at the time of writing comprise the following:

   I GIS and Generalization: Methodology and Practice   edited by J-C. MULLER, J-P. LAGRANGE and R. WEIBEL

   II Geographic Objects with Indeterminate Boundaries   edited by P. BURROUGH and A. FRANK

   IV Spatial Analytical Perspectives on GIS   edited by M. FISCHER, H. SCHOLTEN and D. UNWIN

For these reasons the work described in the series is likely to be of considerable importance in the context of the growing European GIS community. However, given that GIS is essentially a global technology most of the issues discussed in these volumes have their counterparts in research in other parts of the world. In fact there is already a strong US dimension to the GISDATA programme as a result of the collaborative links that have been established with the National Center for Geographic Information and Analysis through the National Science Foundation. As a result it is felt that the subject matter contained in these volumes will make a significant contribution to global as well as European debates on geographic information systems research.

                                                              Ian Masser and *François Salgé*

Editors' Preface

This book presents and analyses the findings of research on the diffusion of GIS in local government in nine European countries. It is an outcome of the first specialist meeting of the European Science Foundation GISDATA programme held in Knutsford, England, in October 1993. GISDATA is a 4-year scientific programme on key research topics relating to European-wide data integration, database design and socio-economic and environmental applications. It seemed therefore natural to start the programme by focusing on the extent of diffusion of GIS in Europe as this provides an essential background to subsequent meetings on data availability in Europe (July, 1994) and geographic information and the information society (Spring, 1996) as well as an important insight into current discussions at the European level on the establishment of a geographic information infrastructure. The focus on local government is particularly useful as this sector is one of the largest users of GIS. It also includes a wide range of applications carried out by different professional groups, and offers the opportunity of studying the extent to which the diffusion of innovations like GIS is sensitive to national issues like the institutional set-up and the availability of data.

The meeting in Knutsford was organized by a small task-force chaired by Ian Masser (Sheffield) and including Michael Wegener (Dortmund), Antonio Arnaud (Lisbon), Heather Campbell (Sheffield), Massimo Craglia (GISDATA Co-ordinator) and Hans Kiib (Aalborg). During the planning stage, it was recognized that very few systematic studies had been carried out on the diffusion of GIS in different European countries, and that the field was largely characterized by personal accounts of individual experiences or by partial studies with no comparable methodology. Against this background, the experience built up at Sheffield over the previous 5 years provided a useful starting point for a more systematic analysis of the diffusion of GIS in local government. This was due to the multiplier effect of a research project on the impact of GIS on British local government by Heather Campbell and Ian Masser which had stimulated other studies such as Massimo Craglia's work on GIS in Italian local government and Dimitris Assimakopoulos's study of the Greek GIS community. During the planning period before the specialist meeting Hans Kiib also spent 6 months at Sheffield adapting the survey methodology developed in the British study for his own research in GIS in Danish local government. At the same time Heather Campbell made a brief visit to Portugal to study GIS in Portuguese local government and a Portuguese colleague Lia Vasconcelos visited Sheffield to study GIS in British local government under the British Council Treaty of Windsor scheme.

Under these circumstances it is not surprising to find that there is a distinct Sheffield flavour to this book in that seven out of the thirteen chapters are authored or co-authored by current members of the Department of Town and Regional Planning. It is also worth noting that much of the work that is described in the book is the product of research carried out since the Knutsford meeting. This accounts for the delay in completing the manuscript. We are however much in debt to all the participants in the Knutsford meeting and to the other contributors to the book for their commitment and dedication in finding the time and financial resources to carry out this research. Without them this book would have never seen the light of day.

The editors are also particularly grateful to Michael Wegener for his constant inspiration and helpful suggestions and to Dale Shaw for her unflappable patience in typing and formatting the manuscript.

Ian Masser
Heather Campbell
Massimo Craglia

Contributors

Antonio Morais Arnaud
Faculdade de Ciencas e Tecnologia, Universidade Nova de Lisboa, Quinta da Torre, P-2825 Monte de Caparica, Portugal

Dimitris Assimakopoulos
Department of Town & Regional Planning, University of Sheffield, Sheffield S10 2TN, UK

Malgorzata Bartnicka
Polska Akademia Nauk, Instytut Geografii i Przestrzennego, Zagospodarownia, ul. Krakowskie Przedmiescie 30, 00-927 Warsaw, Poland

Slavomir Bartnicki
Polska Akademia Nauk, Instytut Geografii i Przestrzennego, Zagospodarownia, ul. Krakowskie Przedmiescie 30, 00-927 Warsaw, Poland

Heather Campbell
Department of Town & Regional Planning, University of Sheffield, Sheffield S10 2TN, UK

Luisella Ciancarella
ENEA-AMB. MON. AMOCS, Viale GB Ercolani 8, 40138 Bologna, Italy

Massimo Craglia
Department of Town & Regional Planning, University of Sheffield, Sheffield S10 2TN, UK

Joao D. Geirinhas
Instituto Nacional De Estatistica, Lisbon, Portugal

A. Graafland
Delft University of Technology, Thijsseweg 11, PO Box 5030, 2600 GA Delft, The Netherlands

Hartwig Junius
Fachbereich Raumplannung, University of Dortmund, Postfach 500500, D-44221 Dortmund, Germany

Hans Kiib
Department of Development & Planning, Aalborg University, Fibigerstraede 11, 9220 Aalborg st, Denmark

Marek Kupiszewski
School of Geography, University of Leeds, Leeds LS2 9JT, UK

Ian Masser
Department of Town & Regional Planning, University of Sheffield, Sheffield S10 2TN, UK

Philippe Miellet
TED-Alitech, 17 Rue Abbé de l'Epée, 3400 Montpellier, France

Enzo Ravaglia
Istituto di Architettura ed Urbanistica, Università di Bologna-Facoltà di Ingegneria, Viale Risorgimento 2, 40136 Bologna, Italy

Piero Secondini
Istituto di Architettura ed Urbanistica, Università di Bologna-Facoltà di Ingegneria, Viale Risorgimento 2, 40136 Bologna, Italy

Edi Valpreda
ENEA-AMB. MON. AMOCS, Viale GB Ercolani 8, 40138 Bologna, Italy

Lia Vasconcelos
Faculdade de Ciencas e Tecnologia, Universidade Nova de Lisboa, Quinta da Torre, P-2825 Monte de Caparica, Portugal

Michael Wegener
Institut fur Raumplannung, University of Dortmund, Postfach 500500, D-44221 Dortmund, Germany

Notes on Contributors

Antonio Morais Arnaud is a Research Scientist at the Computer Science Department, New University of Lisbon (UNL) and is invited Professor at the Higher Institute for Statistics and Information Management (ISEGI). With a Statistics and Operations Research background, he has been involved in a number of projects on data availability for local planning since 1980, namely Census small area statistics (Municenso), fiscal, electoral and administrative statistics, having published a number of papers in local statistics. He led several pilot projects in co-operation with Municipalities and has been consultant at the National Institute of Statistics (INE) for the implementation of a small area statistics programme Munistat. He chaired the 13th Urban Data Management Symposium in May 1989, Lisbon, and is a founder member of the Urban Data Management Society. He has been a member of the core team which launched the ESF GISDATA programme.

Dimitris Assimakopoulos was educated as a civil engineer at the University of Patras, Greece, where he also worked for 2 years under Professor Nicos Polydorides for the Ursa-Net project. Since October 1992 he has been a research associate at the Department of Town and Regional Planning, University of Sheffield, England, where he is completing a PhD on the development of the Greek GIS community, under the supervision of Professor Ian Masser.

Malgorzata Bartnicka and **Slavomir Bartnicki** are the owners of IMAGIS, a successful GIS consultancy company in Poland

Heather Campbell is Lecturer in Town & Regional Planning at the Department of Town & Regional Planning at the University of Sheffield. Research interests centre on the organizational and institutional issues influencing the effective implementation of computer technology in planning practice, policy innovation, and the changing nature of urban management. She has been invited to participate in NATO, European Science Foundation and US National Science Foundation Workshops and has been a visiting Research Fellow at the State University of New York at Buffalo.

Luisella Ciancarella is Senior Researcher in the environmental field at the Italian National Agency for New Technologies, Energy and Environment (ENEA) where she has been working since 1984. Her research interests include environmental planning methods, the environmental impact assessment of human activities and the application of Geographic Information Systems to these topics. Since 1992, she has been a member of an Italian working group for the analysis of GIS diffusion and applications in Public Administrations and has published widely in this field.

Massimo Craglia Lectures on Information Management and GIS in Planning at the Department of Town & Regional Planning of the University of Sheffield, and is the research co-ordinator of the European Science Foundation GISDATA programme. His research interests include comparative planning and GIS applications in town and regional planning.

Ad Graafland is a Lecturer in GIS/LIS at the Department of Geodesy of the University of Technology in Delft. He has been working there since 1986 as well as collaborating with local government and private consultancies in the fields of traffic and town planning. His research interests include the organizational aspects of GIS/LIS, information analysis and information policy planning. In 1993 he received his doctorate on the subject: Geo-information systems in Dutch Municipalities and has published extensively on this subject.

Joao D. Geirinhas is an Environmental Engineer currently working at the National Institute of Statistics (INE), where he has been involved, among others, in the development of a CD-ROM (Census Portugal) in its geographical component. He is also the President of the Portuguese GIS Users Association and the Executive Editor of BIG, the Geographic Information Bulletin. Two years ago he organized the ESIG (National Meeting of GIS Users) 93 and presently he is launching the ESIG 95. He has been working also as a consultant, namely for the Luso-American Foundation for Development. His interests focus mostly on Urban and Regional Planning, Geographic Information Management and Geodemographics.

Hartwig Junius teaches GIS, remote sensing and surveying at the Department of Spatial Planning at the University of Dortmund and is head of the department's GIS laboratory. His research interests include planning cartography and GIS. He has been active in several commissions on spatial information management in local governments in Germany and has published numerous articles on GIS.

Hans Kiib is Associate Professor at the Department of Development and Planning, Aalborg University, Denmark. He has been member of the GISDATA Task Force Group on Diffusion and his research interests include strategic planning with particular interest in planning methods, planning tools and GIS, integrated information systems in local government and media integrated GIS tools and methods in town planning and urban renewal.

Marek Kupiszewski is a Lecturer in Geography at the University of Leeds. He previously was a Research Fellow and Head of the GIS Laboratory at the Institute of Geography and Spatial Organisation of the Polish Academy of Sciences in Warsaw. His research concentrates on the problems of research dissemination and population geography and demography in Central and Eastern Europe. He has published over 50 monographs and contributions to journals and books.

Ian Masser is Professor of Town & Regional Planning at the University of Sheffield, and co-director of the European Science Foundation GISDATA programme. His research interests include geographic information management, planning methods and comparative planning. He is author/editor of 13 books on these topics and has published over 200 contributions to books and journals.

Philippe Miellet has a Doctorate in Geography and for 5 years was an active member of the RECLUS GIS research team in Montpellier. More recently he has moved to the private sector in a company developing GIS applications for land use planning and development studies. Philippe's research interests include both GIS applications for local authorities and GIS teaching methods. He is co-author of Géocube, an interactive GIS encyclopaedia.

Enzo Ravaglia is a young researcher working with the University of Bologna and the Italian National Agency for New Technologies, Energy and Environment (ENEA). His research interests include spatial analysis of socio-economic phenomena with GIS and integration of topographic data for local government GIS applications.

Piero Secondini, since 1978 has taught 'Techniques of urban and regional analysis' (since 1982 as Associate Professor) at the Institute of Architecture and Urban Planning of Bologna University. He is author (or co-author) of many publications in this field (8 books and 50 articles). He is a member of AISRe (Associazione Italiana di Scienze Regionali) and of URISA; he also acts as a consultant to the private and public sector in the field of regional, environmental and retail networks planning.

Michael Tabeling graduated from the Department of Spatial Planning at the University of Dortmund and presently is with the consultancy of Ferdinand and Ehlers in Itzehoe, Germany, working in the field of environmental assessment.

Edi Valpreda is a Senior Researcher in the environmental field at the Italian National Agency for New Technologies, Energy and Environment (ENEA) in Bologna, where she has worked since 1986. Her research interests include quaternary geological studies, human induced subsidence phenomena, recent evolution of shore lines and applications of GIS to these topics including the integration between GIS and Geostatistics. Since 1992 she has been a member of an Italian research group on GIS diffusion and applications in the Italian Public Administration and has published in major national and international journals.

Lia T. Vasconcelos is a Lecturer in the Department of Environmental Engineering of the New University of Lisbon. Most of her work has focused on the use of information and new technologies, in particular within the context of environmental quality and growth management, namely in areas related to spatial organization and analysis.

Michael Wegener is with the Institute of Spatial Planning and teaches urban and regional theory and planning theory at the Department of Spatial Planning at the University of Dortmund. His research interests include issues of urban development both at the European and local scale and GIS-based spatial models. He has written extensively on these topics in several books and numerous journal articles.

The **European Science Foundation** is an association of its 55 member research councils, academies, and institutions devoted to basic scientific research in 20 countries. The ESF assists its Member Organizations in two main ways: by bringing scientists together in its Scientific Programmes, Networks and European Research Conferences, to work on topics of common concern: and through the joint study of issues of strategic importance in European science policy.

The scientific work sponsored by ESF includes basic research in the natural and technical sciences, the medical and biosciences, and the humanities and social sciences.

The ESF maintains close relations with other scientific institutions within and outside Europe. By its activities, ESF adds value by cooperation and coordination across national frontiers and endeavours, offers expert scientific advice on strategic issues, and provides the European forum for fundamental science.

This volume is the first of a new series arising from the work of the ESF Scientific Programme on Geographic Information Systems: Data Integration and Database Design (GISDATA). This 4-year programme was launched in January 1993 and through its activities has stimulated a number of successful collaborations among GIS researchers across Europe.

Further information on the ESF activities in general can be obtained from:
European Science Foundation
1 quai Lezay Marnesia
67080 Strasbourg Cedex
tel: +33 88 76 71 00
fax: +33 88 37 05 32

Dr Geir-Harald Strand
Norwegian Institute of Land Inventory
Box 115, N-1430 Ås
Norway

Dr Antonio Susanna
ENEA DISP-ARA
Via Vitaliano Brancati 48
00144 Roma
Italy

CHAPTER 1

Introduction

MASSIMO CRAGLIA and IAN MASSER

Geographic information systems (GIS) are a multimillion dollar industry which is growing very rapidly at the present time. Estimates of its size vary according to the measures used. Payne (1993), for example, estimates that the core European GIS market was worth 415 million ECU (US$ 515 million) by comparison with an American market of $466 million in 1991. He also estimates that the European market had grown at a compound annual rate of 14% until 1993 reaching some 539 million ECU.

These estimates relate only to core revenue which is a relatively small part of the total market when measured in terms of end user expenditure. Once the total expenditure on hardware, software and services including those bought from third parties is taken into account Payne estimates the total value of the European GIS market to be worth around 4 billion ECU in 1993.

These estimates do not include the costs to government associated with the collection and management of geographic information. A recent report for the European Commission (1995) estimates that government spending on geographic information alone accounts for around 0.1% of gross national product or about 6 billion ECU for the European Union as a whole. Similarly, in the United States the Office of Management and Budget has estimated that federal agencies alone spent $4 billion annually to collect and manage domestic geospatial data (Federal Geographic Data Committee, 1994, p. 2).

The growth of the core GIS market shows little sign of slowing down. Dataquest (Gartzen 1995), for example, estimates that revenue from GIS software world-wide is likely to grow at an average annual rate of 13% up to 1998, and that in the short term North America and Europe will continue to dominate the world GIS market.

The scale of the GIS industry makes it necessary to study the diffusion of this technology and its impacts on society as a whole. This is because the technology has the potential to change fundamentally the use of geographic information in a wide range of circumstances and it must be recognized that not all of these changes will be beneficial. The largest users of GIS technologies are central and local government agencies although the number of private sector applications has substantially increased in the last few years. In countries such as Germany and Great Britain, local government applications probably account for somewhere between a quarter and a third of the total market.

Given the size of the GIS industry and the extent to which it is dominated by a relatively small number of large users, it is surprising to find that relatively little systematic research has been carried out either on the development of the geographic information services industry itself or on the diffusion of geographic information technologies in key sectors such as local government. Local government is a particularly interesting field for diffusion research because it covers a wide range of applications in a great diversity of settings. There is also a strong national dimension to local government which is reflected in the expectations that people have with respect to local authorities, the tasks that they carry out and the professional cultures that have come into being to support them.

This book explores the diffusion of GIS in local government with particular reference to the experience of nine European countries. The term 'diffusion' refers to the process whereby technological innovations such as GIS are adopted and taken up by various user groups. According to Rogers (1993, p. 15) there are four main elements in the diffusion of new ideas '(1) an innovation, (2) which is communicated through certain channels, (3) over time, (4) among members of a social system'. This definition highlights the extent to which the outcome of the diffusion process is dependent not merely on the technical quality of the innovation itself but also on the social and political context within which it takes place.

There is a great deal of evidence from research on the diffusion of technological innovations which suggests that the process over time can typically be represented by an S-shaped or logistic curve (see, for example, Rogers, 1993). In essence this implies that the initial rate of adoption of technological innovations is likely to be relatively slow until a critical mass of users is achieved after which the rate of adoption increases very rapidly and the direction of the curve moves away from the horizontal axis towards the vertical axis. As time progresses and levels of adoption near saturation the curve tapers off again towards the horizontal axis.

There are also two basic spatial models of the spatial dimension of the diffusion of technological innovations (see, for example, Hägerstrand, 1967) which are of interest. The first of these is the hierarchical model which postulates that adoption will begin in the larger centres/organizations and subsequently diffuse to smaller centres/organizations. Large centres are seen as being more open to the outside world because of their size and more likely to take on the role of pioneers because of the resources at their disposal. Conversely it is claimed that small centres are less likely to be aware of new developments and also less likely to have the necessary resources to experiment with new tools. The second spatial diffusion model is the core-periphery model. Like the hierarchical model this assumes that adoption will begin in core cities or regions because of their size and their links to the outside world and then spread outwards to peripheral cities or regions which are less cosmopolitan in character and have fewer resources at their disposal.

The book builds on the findings of the NATO Advanced Research Workshop on 'Modelling diffusion and use of geographic information technologies' which took place at Sounion in Greece from 8 to 11 March 1992 (Masser and Onsrud, 1993). This addressed the following questions:

1. What can GIS researchers learn from research elsewhere on organizational behaviour and technological innovation?

2. What are the main features of the GIS diffusion process?

3. To what extent is the adoption and utilization of GIS facilitated or impeded by the institutional and organizational context within which it takes place?

4. What mechanisms might be used at the regional, national and international levels to facilitate the diffusion of GIS?

The present book is primarily concerned with exploring the second and third questions in much greater depth than was possible during the NATO workshop. It examines the experiences of GIS diffusion in the context of local government in nine European countries from a cross national comparative standpoint. Of particular importance in this context is the extent to which the nine case studies draw upon a common methodology based on the research on this topic that has been carried out at Sheffield over the last 5 years. This includes both survey and case study research on British local government (see, for example, Campbell and Masser, 1992, 1995; Masser and Campbell, 1993) as well as comparative research on GIS in Britain and other European countries (see, for example, Campbell and Craglia, 1992; Craglia, 1994).

Another important feature of the book is the way in which it also addresses some of the broader theoretical issues involved in GIS diffusion research. This is particularly evident in the contributions by Michael Wegener and Ian Masser and by Heather Campbell in the first part of this book which directly consider two of the most important limitations of diffusion research as regards GIS that were identified by Everett Rogers in his keynote address to the NATO Advanced Research Workshop in Greece. The first of these concerns the extent to which there is an implicit pro innovation bias in much diffusion research. In Rogers' opinion

> this lack of recognition of the pro innovation bias makes it especially troublesome, and potentially dangerous in an intellectual sense. This bias leads diffusion researchers to ignore the study of ignorance about innovations, to underemphasise the rejection or discontinuance of innovations, to overlook re-invention, and to fail to study anti diffusion programmes designed to prevent the diffusion of 'bad' innovations (like drugs, or cigarette consumption, or aircraft skyjacking, for example)   (Rogers, 1993, p. 19).

With this in mind Michael Wegener and Ian Masser consider some potential social and political consequences of GIS diffusion in the context of four contrasting scenarios of future Brave New GIS Worlds. The benchmark scenario is a trend scenario which represents what may happen as a result of the incremental diffusion of GIS according to past trends. It is largely benign in nature reflecting the pro innovation bias in most of the GIS literature. This is not the case with the other three scenarios. The first of these is a Market scenario which extends current tendencies towards the commodification of information and restricts access to those who have power. The second Big Brother scenario highlights the potential of GIS for surveillance and control in all aspects of life, while the Beyond GIS scenario shows how GIS could also be used to promote wider democratic debate and facilitate grass roots empowerment.

The basic objective of this contribution is to counter the pro innovation bias in most of the GIS literature and demonstrate that GIS diffusion has negative as well as positive impacts. Heather Campbell's contribution focuses on the other limitation of most diffusion research when seen from the standpoint of GIS: its emphasis on individuals rather than organizations. As Rogers points out, studying the innovation process in organizational settings such as local government necessitates a shift in the classic diffusion model. In this context the adoption of a new technology by an organization is only part of the diffusion process as it is only considered to be successful if it leads to its implementation and utilization in practice. In this respect,

> the individual plays an active creative role in the innovation process by matching the innovation with a perceived organizational problem and perhaps in re-inventing the innovation. An innovation should not be conceived as a fixed, invariant and static element of the innovation process, but as a flexible and adaptable idea that is consecutively defined and re-defined as the innovation process gradually unfolds
>
> (Rogers, 1993, pp. 19–20).

With these considerations in mind Campbell examines three explanatory theories of organizational change with respect to the diffusion of GIS: technological determinism which assumes that innovations diffuse simply because of their inherent technological advantages over existing practice, economic determinism which regards computerization as an essential prerequisite for economic survival both in the public and private sectors, and the social interactionist approach which assumes that technology is socially constructed and views the diffusion of innovations in terms of the interaction between the technology and potential users within a particular cultural and organizational context. Her analysis highlights the weaknesses in the two deterministic approaches and the need to develop research strategies based on the social interactionist approach in further diffusion research on GIS in organizations.

Campbell also makes the point that differences in the extent of diffusion are not exclusively related to particular national circumstances and that there are also often striking similarities between organizations located in widely varying institutional contexts. Notwithstanding this, the second and third parts of the book present the findings of nine national case studies of the diffusion of GIS in local government within Europe. The essential difference between the second and third part is with respect to the methodology used by the authors.

The five case studies that form the second part of the book make use of a similar survey methodology whereas the four case studies in the third part present a number of alternative perspectives of the diffusion process in local government in particular countries. A common reference point for the first five case studies was the survey of the diffusion of GIS in British local government carried out at Sheffield during the first half of 1991 (Campbell and Masser, 1992). This involved a comprehensive telephone survey of all 514 local authorities in Great Britain to assess the geographic extent of diffusion, the nature of systems that were in place, the organizational arrangements that have been made within different authorities and the perceived benefits and problems associated with the acquisition of GIS.

A number of adaptations of this methodology have been made in the five case studies to meet local circumstances. Masser and Campbell's analysis of GIS diffusion in British local government describes the findings of a second survey carried out during the summer of 1993 using the same methodology as that used in the earlier survey. As a result it is possible to make direct comparisons between the two sets of survey findings which highlight the volatile nature of GIS in local government and illustrate the dynamics of the diffusion processes that have taken place. In the next chapter Hartwig Junius and his colleagues discuss developments in Germany with particular reference to land information systems. For practical reasons their detailed analysis is limited to the 86 German cities with populations of over 100 000. Luisa

Ciancerella and her colleagues explore the diffusion of GIS in Italy at a time of political upheaval when considerable changes are taking place to the traditional cultures of both central and local government. Their analysis is based primarily on the findings of comprehensive surveys of GIS at the regional and provincial levels of local government. In the fourth chapter in this part, Antonio Arnaud and his colleagues present an interesting account of GIS at the point of very rapid take off in Portugal as a result of a series of proactive government initiatives to modernize local government. This draws largely on the findings of a postal survey of municipalities carried out by the authors. Finally the diffusion of GIS in one of the most decentralized countries in Europe with respect to the powers that have been devolved to local government is described by Hans Kiib. His analysis is based on the findings of a comprehensive survey of Danish counties and municipalities.

The four case studies contained in the third part of the book complement those of the previous section by bringing a number of fresh perspectives to the analysis of GIS diffusion. Dimitris Assimakopoulos examines the structure of the technological community that has grown up in Greece as a result of GIS diffusion. This consists of a core of about 60 members working in a variety of administrative settings including local government. He also presents the findings of two case studies of local government which provide interesting examples of innovative organizational cultures that have sought to exploit the opportunities opened up by GIS. Philippe Miellet's account of GIS in French local government looks at GIS diffusion from a historical perspective. It shows the extent to which recent developments have built upon the long tradition of geographic information handling in many of France's larger cities. In the next chapter Malgorzata Bartnicka and her colleagues provide some insights into diffusion strategies for GIS in what was until recently a centrally planned economy in Poland. Under these circumstances, the context of GIS diffusion in local government as elsewhere in Poland, is one of rapid change in which old and new exist and fight for supremacy. In the final chapter of this section Ad Graafland provides a broad overview of the diffusion of GIS and graphical automated systems among Dutch municipalities. Against this background he develops a four stage model of GIS diffusion in local government from initiation, to local control, then to infrastructure development and finally to integration.

In the last section of the book Ian Masser and Massimo Craglia undertake a comparative evaluation of GIS diffusion in local government in the nine countries whose experiences have been described in the second and third parts of the book. The process of comparative evaluation is divided into three stages. The first of these involves the construction of a profile for each country which summarizes the distinctive features of each national experience. These features are then compared with those of the other eight countries within a common analytical framework while the third and final stage explores the extent to which a typology of GIS diffusion experience can be developed on the basis of the main differences between countries.

Taken as a whole the contributions to this book contain both reflections on the nature of diffusion processes which involve new technologies such as GIS as well as detailed case studies of national experiences with respect to the diffusion of GIS in local government. The findings of these case studies make it possible to carry out a more systematic comparative evaluation of these experiences than has hitherto been the case. In this way it is felt that the book makes a useful contribution to diffusion theory as well as to substantive knowledge about the current state of diffusion in one of the key stakeholders in the GIS community.

REFERENCES

CAMPBELL, H. and CRAGLIA, M., 1992. The diffusion and impact of GIS on local government in Europe: the need for a European wide research agenda, *Proceedings of the 15th European Urban Data Management Symposium*, Lyon 16–20 November, pp. 133–55, Delft: Urban Data Management Society.

CAMPBELL, H. and MASSER, I., 1992. GIS in local government: some findings from Great Britain, *International Journal of Geographic Information Systems*, **6**, 529–46.

CAMPBELL, H. and MASSER, I., 1995. *GIS and Organisations*, London: Taylor and Francis.

CRAGLIA, M., 1994. Geographic information systems in Italian Municipalities, *Computers Environment and Urban Systems*, **18**, 381–475.

EUROPEAN COMMISSION, 1995. *GI 2000: towards a European Geographic Information Infrastructure*, Luxembourg, DGXIII.

FEDERAL GEOGRAPHIC DATA COMMITTEE, 1994. The 1994 plan for the National Spatial Data Infrastructure: building the foundations of an information based society, Reston: Federal Geographic Data Committee.

GARTZEN, P., 1995. GIS in an ever changing market environment, *Proceedings of the Joint European Conference and Exhibition on Geographic Information*, pp. 433–6, Basel: AKM Messen AG.

HÄGERSTRAND, T., 1967. *Innovation Diffusion as a Spatial Process*, Chicago: University of Chicago Press.

MASSER, I. and CAMPBELL, H., 1993. The impact of GIS on local government in Britain, in Mather, P.M. (Ed.) *Geographic Information Handling: Research and Applications*, pp. 273–86, Chichester: John Wiley.

MASSER, I. and ONSRUD, H.J. (Eds), 1993. *Diffusion and Use of Geographic Information Technologies*, Dordrecht: Kluwer.

PAYNE, D.W., 1993. GIS markets in Europe, *GIS Europe*, **2**(10), 20–2.

ROGERS, E.M., 1993. The diffusion of innovations model, in Masser, I. and Onsrud, H.J. (Eds), *Diffusion and Use of Geographic Information Technologies*, pp. 9–24, Dordrecht: Kluwer.

PART I

Reflections on GIS Diffusion Research

CHAPTER 2

Brave New GIS Worlds

MICHAEL WEGENER and IAN MASSER

INTRODUCTION

Traditionally diffusion research has a pro-innovation bias. Diffusion is studied as a process by which older, outdated technologies are replaced by more advanced, more efficient and hence more beneficial ways of doing things. The first systematic studies of diffusion processes by Rogers (1962) and Hägerstrand (1967) looked into the way new technologies, such as new farming techniques, the telephone or television, gradually penetrated their markets. The implicit assumption, never discussed, was that adoption of the new technology was in the interest of the adopters.

However, the view that a new technology is always better than the older one it displaces has long been discredited by the dialectic of technological progress—the experience that more often than not a successful technology, once it has become dominant, also displays a destructive dark side. At least in the case of farming, this dialectic has become commonplace. The 'green revolution' helped farmers to multiply their crops by introducing fast-growing plants with higher yields, more efficient fertilizers and more effective pesticides, yet also brought new risks of water contamination and soil erosion, contributed to rural unemployment and depopulation and caused grave imbalances in global food markets.

Diffusion research was not able to visualize these long-term impacts because it was concerned with the early phases of diffusion, in which the new technology was still virgin and innocent. As only the beneficial aspects of the new technology were seen, rapid diffusion was interpreted as success and lack of it as deplorable backwardness. In fact the most frequent motivation for diffusion research has been to identify barriers to the rapid adoption of the new technology and, once these have been identified, to recommend strategies to overcome them (cf. Rogers, 1993).

Diffusion research concerned with the adoption of geographic information systems (GIS) is no exception. The growing volume of studies on the adoption and use of GIS in the United States and in European countries have been stimulated not by academic interest but by the well-intended drive to identify and help remove institutional and technical bottlenecks to their universal distribution and application —not surprisingly, because the authors of the studies generally are members of the GIS community, i.e. individuals with a strong interest in this rapidly growing market

9

(see, for instance, Masser and Onsrud, 1993). The dangers of a pro-innovation bias of this kind of research are obvious. Diffusion research which only sees the positive side of the new technology it is concerned with is unable to distinguish between backwardness and other more serious reasons for differences in speed or intensity of adoption. Even where it subsumes barriers to adoption under the broad but diffuse heading of 'cultural factors', it will always view these as regrettable and something to overcome, and this will inevitably colour the conclusions drawn from the research.

Yet there are good reasons to move beyond a naive all-out promotion of GIS and arrive at a differentiated and balanced stance which carefully weighs their obvious benefits against their potential risks. As long as there have been computers, there have been warnings that the information revolution may endanger fundamental human values. The 'data bank' and 'information system' have always carried the Janus-face of unlimited knowledge about and control over the individual. There have always been fears that this kind of knowledge, in the hands of irresponsible bureaucrats, law-and-order police officers, power-hungry politicians, criminal organ- izations or unpredictable fanatics, could be used to undermine democracy and individual privacy. However, the recent success of geographic information systems has given these warnings a new dimension. Geographic information systems, with their capability to localize every conceivable object or activity in Cartesian space, are the ultimate expression of the rationalist dream of measuring and knowing *everything*. In combination with concurrent technologies such as electronic data interchange, GIS introduces a new powerful threat, that of total *spatial* control.

The first concerns that GIS might be far more dangerous than previous information systems have been expressed not by proponents of GIS, but by social and political scientists and a few critical geographers (e.g. Smith, 1992; Obermeyer, 1992; Curry, 1993; Lake, 1993, Onsrud *et al.*, 1994; Pickles, 1994). The GIS community itself has largely remained confined to an uncritical promotional attitude towards GIS. GIS journals such as *GIS Europe* are technology- and application-oriented and rarely deal with the social impacts of GIS. Academic discussions orbit around epistemological or methodological issues of GIS or what GIS do to geography (e.g. Openshaw, 1991, 1992, 1993; Taylor and Overton, 1991; Couclelis, 1993) but hardly touch upon their limits and risks. However it is time that GIS experts also become aware of the debate on the impacts of GIS on society and develop adequate answers to its serious questions. This chapter tries to contribute to this debate by suggesting a number of scenarios of possible future GIS diffusion which capture the range of perceptions of the impact of GIS on society found in different countries of Europe today.

GIS TODAY

Geographic information systems (GIS) include a wide range of different applications including automated mapping and facilities management as well as land information systems. As the number of applications grows, the term GIS is used increasingly as shorthand for a great diversity of computer-based applications involving the capture, manipulation, analysis and display of geographic information and the associated services that go with them.

Although many of the basic concepts underlying GIS were developed more than 20 years ago, the computer technology required to manage large amounts of geographic information and display them in graphical form has only been available since the mid-1980s. Since that time the GIS hardware and software industry has dramatically expanded in terms of both the number and range of applications. It is estimated that sales of GIS hardware since 1985 have grown at rates well over 10% per annum, while software sales have increased by 15–20% each year. As a result the volume of hardware sales has doubled every 6 six years since 1985, while that of software has doubled every 3–4 years.

The pace of technological innovation is still accelerating and the range of applications continues to expand. In fact the number of GIS facilities in operation has grown at an even faster rate than overall sales as an increasing number of budget-price installations come on the market.

The main users of GIS are central and local government agencies and the utility companies. Together these account for well over half the overall GIS market in most countries. Other important application areas are in the field of environmental management and facilities planning. Over the last few years there has been a considerable increase in the number of business applications for sales analysis and marketing. These already account for 8% of the GIS market and it is forecast that their share will rise to at least 15% over the next 5 years. Other potential fields which are still to be exploited include the use of GIS in vehicle navigation systems.

The utilization of geographic information systems is heavily dependent on the availability of digital topographic data. As a result of variations in national government policies towards data provision, there are considerable differences between countries in terms of the availability of digital topographic data at both the large and small scales. There are also important differences in the cost of information of this kind to users, as the providers in some countries attempt to recover the cost of data provisions. In Britain, for example Ordnance Survey data are protected by copyright, and the agency itself already recovers 70% of its costs, whereas the TIGER files developed by the United States Bureau of the Census are available at minimal cost without copyright restrictions.

The growth of GIS over the last years has stimulated a massive growth in specialist GIS services of all kinds. These range from bureaus specializing in digitizing, automated data capture and customizing spatial data bases to management consultancies advising agencies on the benchmarking and implementation of particular GIS packages. An important sub-group of GIS services is associated with the development of customized software for particular application fields. Of particular importance in this respect is the development of decision support systems for commercial marketing operations. As the number of systems in operation has increased, there has been a parallel growth in legal actions regarding the accuracy and reliability of the information provided by them. As a result, litigation and claims for liability and compensation are emerging as an important growth area.

The spatial impacts of these developments are not homogeneous. The GIS market is particularly well developed in North America, whereas the European GIS market is still divided by national interests and highly fragmented in character. There is little agreement on common standards, and there are considerable differences in the professional cultures that are involved in GIS applications. Generally the north and west European countries have experienced higher levels of GIS penetration than those of southern and eastern Europe. However, there are also marked variations within countries in the level of GIS penetration, particularly in southern and eastern Europe.

THE FOUR SCENARIOS

This is the situation from which the four scenarios start. Each of them is a projection of one possible evolution of the uses of GIS and their impact on society. The first scenario is the *Trend* scenario characterized by incremental diffusion of information systems along the lines experienced in the past. The other three scenarios highlight and exaggerate specific tendencies that can be observed today. The *Market* scenario extends current tendencies towards commodification of information, which restricts access to information to the more powerful. The *Big-Brother* scenario dramatizes the potential of GIS to be used for surveillance and control by fully integrated omniscient systems, which pervade all aspects of life. The *Beyond-GIS* scenario, finally, speculates on how information in the public domain might contribute to more democratization and grassroots empowerment. All four scenarios look 20 years into the future and are expressed as narratives of a person looking back to the 1990s.

The Trend Scenario

The year is 2015. The past 20 years have been a period of stupendous technological developments. All of them have been based on innovations made in the 1980s, but nobody at that time would have expected the speed by which they have penetrated their markets. New materials have brought unprecedented levels of miniaturization, memory and computing speed of all kinds of electronic devices. Telecommunications, cable and computer companies have merged into transnational media conglomerates. Fibre optics, cable, cellular radiophony and satellite communications have grown together into an integrated multi-layer network of information superhighways bringing fax, e-mail, smart TV and electronic data interchange to every home and office. Artificial intelligence, multimedia and virtual reality have amalgamated to create new kinds of computer applications that are more user-friendly, entertaining and unobtrusive than ever.

   All these developments have had their impacts on GIS. As a result of the immense advances in performance and reduction in cost of both hardware and software, the number of GIS installations, the development of user-friendly interfaces and the range of applications have multiplied to the extent that geographic information systems are now used universally like spreadsheets and data base management software before them. One impact of universal GIS is that most users make use of GIS facilities without ever being aware that they are doing so. This is particularly the case with multimedia applications using virtual-reality GIS. Together with GIS installations, the number, size and diversity of spatial data bases has grown explosively. Today there is a huge variety of public and private spatial data bases for all conceivable purposes, from postcode systems precise to the letter box to multimedia, virtual reality house catalogues or travel guides. GIS education is now part of the conventional school curriculum. The GIS industry has become increasingly specialized and fragmented in order to meet the great diversity of demands placed on it by different applications groups. The term GIS is used less frequently than during the 1990s, and when it is used, it tends to be prefaced by another term indicating the specific subset of applications that is involved.

   Within Europe as a whole there are still considerable differences between countries.

Although efforts to promote greater harmonization of GIS by the European Organization for Geographic Information have had some success, there are still considerable differences between countries in terms of the data that are collected and the extent to which they are made available to users as well as with respect to the data models and data interchange formats used. Many of the differences between professional cultures also remain, particularly with respect to the key GIS users such as local governments and the utility companies. However, considerable progress has been made in reducing regional disparities within Europe, partly as a result of initiatives of the European Union. As a result, the gap between the European and north-American GIS industries in terms of market penetration has largely disappeared. This is also due to technological developments and automated data capture, which have resolved many of the problems previously faced by information-poor countries.

In 2015, therefore, GIS in Europe is both universal in extent and largely benign in operation, while the applications field as a whole has become highly fragmented and specialized in nature. Variations between European countries still persist despite efforts to promote harmonization. Against this backcloth, the potential of the technology and the capabilities of organizations to manage it are still being constantly tested in practice. Because of the risks involved in such operations, the media contain occasional reports on gross incompetence and inefficiency in public-sector GIS applications as well as about the enormous sums that are being paid out in compensation as a result of court decisions regarding GIS.

The Market Scenario

The year is 2015. The information industry has become the largest and most powerful economic sector. As goods production now largely takes place in the developing countries and in eastern Europe, more than 70% of all economically active persons in western Europe primarily handle information during their daily work. Digital data, text, audio and video, fax, telephone and electronic data interchange have amalgamated into one integrated multimedia information and communications technology. The desktop computer has given way to a flurry of miniaturized, interconnected electronic gadgets from credit card to hand-held super computer. All individuals and households are part of and connected with thousands of electronic networks putting at their disposal all conceivable kinds of deliveries and services. Every transaction in daily life leaves a trace in these networks: orders, sales, invoices, receipts, itineraries, reservations, inquiries, messages, sounds and images.

A large part of the traffic over the networks is geocoded. Every customer or supplier is associated with a unique address, which not only represents a point in geographic space, but is also a node of the transport and telecommunication networks and is linked to a postcode, enumeration district, electoral ward, municipality and county. Attaching a geocode to an item has become so easy that geocoding is used even where it does not serve any other purpose than identifying an object. Every trip, delivery route or electronic message represents a spatial interaction between two addresses and can be aggregated to flows of people, goods and information across the territory. Knowledge about these stocks and flows, about potential customers and the pattern of their activities, is economic power, which can be used to contest or defend a market.

This is why most of the networks and the information they contain is private. In the 1990s, many European governments, following the neoliberal economic doctrine of that time, privatized their postal services, transport and telecommunications networks and enforced a strict cost-recovery policy for government agencies providing post-coded directories or cartographic or statistical services, which traditionally had been free or could be obtained for a nominal charge. Local governments followed suit by privatizing their utility companies and contracted out mapping and surveying tasks. Privacy legislation, which had been overly constraining the information industry in some European countries, was harmonized between European countries in the late 1990s. Today it is legal to collect and trade data on individuals as long as the information appears to be correct.

The result was the emergence of an immense market for value-added telecommunications services and geocoded information. During the 1990s small and medium-sized firms specializing in digital databases with the associated software mushroomed. There was a proliferation of digital road databases for trip planning and fleet management, of small-area population and household databases for marketing planning, and of large-scale digital city maps for real-estate development and property management and sales. Prospective home buyers could browse in virtual reality through offered houses and their environment without actually going there. The same technology was used by travel agencies, instead of bulky catalogues, to market package tours. Other rapidly growing markets were utility planning, facility management and vehicle tracking and navigation systems for the rapidly growing intelligent vehicle-highway systems (IVHS) industry.

Besides these applications for commercial and professional users, a booming market for consumer or home GIS emerged. People could download virtual travel experiences as a surrogate for actual travel to far-away countries, cities or museums— one could even book a trip to ancient Rome. Other popular home applications of GIS were trip planning, geography courses and spatial computer games. As with today's video games, customers were lured into buying cheap hardware to make them captive to expensive software.

It was a period of creative turbulence and confusion. Every conceivable spatial information of commercial value was digitized over and over again by a multitude of data suppliers. Needless to say that all these proprietary databases were of varying accuracy and reliability and incompatible with each other. As competition became fiercer, prices plummeted. There were real data wars between suppliers; even sophisticated encrypting techniques did not prevent massive reverse engineering of data bases resulting in an explosion of litigation about data ownership and copyrights. Because of the proliferation of data suppliers, it became less and less profitable to trade raw data. The real business was to compile customized 'designer information' for the specific purposes of individual clients. As this was more often than not persuasion and manipulation if not deception, the notion of what was 'correct' information underwent a subtle change. As a consequence, litigation on the liability for damages due to the use of incorrect or distorted data emerged as a second fast growing field of legal disputes. In particular some spectacular cases of large-scale fraud in international virtual space created a worldwide legal debate about which country's jurisdiction to apply.

In the late 1990s the market consolidated and many small suppliers of digital geocoded information went out of business. After some spectacular mergers and take-overs, a few big transnational players remained: among them Mitsubishi,

Siemens–Bull and Warner–Murdoch, the US–British media giant, which had ventured into the geoinformation business by swallowing EtakMap of Atlanta, Georgia, and by launching its own fleet of imagery satellites. The Warner–Murdoch (formerly Etak) map encoding system became the factual industry standard. More recently alliances between the geoinformation industry and credit card companies, travel agencies and telecommunication networks operating worldwide have created giant online data banks capable of tracking not only lost luggage but also travellers or customers with any desired detail.

Governments at all levels, once the sole providers of geoinformation, found themselves at the mercy of the information conglomerates. Their retreat from the information scene in the 1990s, based on short-sighted budgetary considerations, proved to be a costly mistake. Since in most European countries now population and employment censuses have been abolished, governments have to pay the market price for the same kind of information which in former times they had produced themselves. Even worse, they do not get all the information available, as certain kinds of data on property values, household income, consumer preferences or travel patterns are too commercially valuable to be released to the public domain. Ironically, the refusal to sell commercially profitable data to government is often justified by reasons of confidentiality, although everybody knows that no such constraints are observed where those data are used for commercial purposes.

The loss of public control over the geoinformation market has seriously affected the status and effectiveness of public planning. Some kinds of data of potential value for local planning have practically disappeared because their collection or updating is not profitable, such as historical or time series small-area data or maps of non-metropolitan areas. Other kinds of data have a negative effect on urban development because they are selectively available only to certain groups. Proprietary information on the socioeconomic composition of neighbourhoods and on property values, for instance, has been used by real-estate agents to speculatively manipulate land and house prices with the effect of displacement of poor households and reinforcement of spatial disparities. In fact, dealing with manipulated real-estate information has become an important field of activity of organized crime.

Other users of geoinformation, who formerly relied on government services for their information needs, are effectively excluded from access today. University research using geographic information is hardly able to afford privately collected data offered at market price. Users without financial means, such as students or citizen groups or protest movements, who need information for their study or political work, have no chance. But nobody complains; people see that information is a commodity and understand that the market does not produce where there is no demand. Nor is the issue discussed in the media which are controlled by the same multi-nationals that produce the data. The free information market is not free for everybody but only for the rich and powerful. The consequence is a widening gap between the information-rich and the information-poor: between those who participate in the information society as providers and manipulators of information and those who participate in it only as consumers and have access only to manipulated information.

The information gap is widening also between regions and countries. Developing countries have become dependent on the transnational geoinformation corporations from whom they buy GIS-processed satellite images indispensable for resource exploration and water supply management. In addition, also east European countries, which had not had the time to develop their own geoinformation industry, are victims of this dependency. After their privatization, the statistical offices and mapping agencies of Poland, the Czech Republic and Hungary were acquired by Mitsubishi, whereas Siemens–Bull succeeded with their bid for designing CISGIS, the distributed geographic information system for the countries of the former Soviet Union.

The Big-Brother Scenario

The year is 2015. What a relief that the opposition of the 1990s against the geoinformation networks had not been successful. The European corporate state depends on reliable intelligence to defend itself against crime and subversive activities. Today it is hard to imagine how the security of residential areas or shopping malls could be guaranteed without efficient spatial surveillance systems. Even driving on highways has become more secure since every vehicle is being monitored by police, although this had been introduced originally merely for accounting purposes.

There had been a time when some people had resented being registered in the new geocoded information systems. There had been even fears that data banks with the capability to track everybody's movements, might endanger basic human rights. Fortunately, these concerns have long been dismissed as exaggerated fabrications of the individualistic liberal period. Today citizens realize that modern information systems are only to their benefit. They appreciate the convenience and safety of the welfare state and are eager that they are correctly represented in as many data banks as possible as home owners, customers, subscribers, patients or drivers and wherever they go, at home or abroad. Of course, people who are denied the privilege of membership may complain; but they must understand that the exclusion of people without credit line is necessary for the protection of the majority. Also people who find their whereabouts tracked in police data banks, such as narcotics dealers, traffic delinquents, HIV positives, homeless, or people with questionable political views may not like this, but they can only blame themselves for being observed in the interest of a safe society.

The integration of the geocoded information systems started in 1998, when Eurostat and Europol, with the help of Siemens–Bull, the European information giant, were amalgamated into the European Intelligence Agency (EIA). It was the task of the new public–private agency to integrate all hitherto isolated national spatial information systems into a coherent hypernetwork of distributed information interchange following the lead of the 'information highways' programmes in the United States and Japan. It was argued that only by this integration would Europe have a chance to compete with these two rivals in the fight for global economic dominance. One can say that the integration of geoinformation did more for the unification of Europe than the Single European Market in 1993.

The impact of this restructuring of the geoinformation scene in Europe was dramatic. It ended the chaos of uncoordinated production of geoinformation by small suppliers of the liberal period. Now it was recognized that spatial information which is freely available to everybody is intrinsically dangerous, whereas in the hands of the corporate state it can guide a society to achieving its highest economic potential. Since the European Freedom-of-Information Act of 1998 therefore every collection of geocoded information has to be licensed by the local subsidiary of the EIA, and any collected geoinformation is classified unless explicitly released by the EIA to the public domain as economically not sensitive. This law has greatly reduced the number of unqualified suppliers of geoinformation and the volume of litigation in this area.

From an engineering point of view, the European information network is a marvellous achievement. DESCARTES (distributed European spatial control and real-time early-warning system) represents the latest advance in hypernetwork technology. It is in fact a network of networks, superseding the hotchpotch of formerly separated and incompatible public (police, secret service) and private (commercial data bank and corporate data and transaction) networks in one grand, unified design—a splendid synthesis between German thoroughness and French elegance. Equipped with latest artificial intelligence, DESCARTES is an adaptive, learning, decentralized system. It has therefore no single primary control centre; its alert rooms are virtually distributed over all its levels, in police headquarters, corporate offices or the various spatial levels of government. In the control rooms planners watch for sensor lights to flash on floor-to-ceiling maps at places where trouble is likely to occur (cf. Wegener, 1987). The EIA has the responsibility of maintaining the network and controlling access to it as well as linking it to similar networks in the United States, Japan and China.

However, DESCARTES was not only an engineering achievement; it also has had a deep influence on the relations between people. Never before had there been such a harmonious society. Violence and street crime have practically disappeared, since all public spaces have been equipped with a video surveillance system; without surveillance people would not feel safe. Most people have asked the authorities to link their homes to the circuits to demonstrate that they have nothing to hide, in fact privacy has become associated with something unethical if not illegal. Surveillance is moulding behaviour in many beneficial ways. For instance, neighbourhoods now look much tidier, since remote sensing has enabled police to monitor garden maintenance.

Of course, where there is much light, there must be some shadow. There remains the misery of those who are excluded from the surveillance society, such as illegal immigrants, tax evaders, or subversive elements living in sewers or abandoned underground tunnels. They do not enjoy the benefits of surveillance but are themselves strictly observed by police and, if necessary, ruthlessly attacked. It remains an interesting question why the authorities have tolerated the existence of this underclass in an otherwise perfect society.

The Beyond-GIS Scenario

The year is 2015. Seen from today, the GIS craze of the 1990s looks like a strange fad. Certainly, there have been some useful applications of geographic information systems in cartography, planning or facility management, but to call this a revolution was a vast exaggeration. More likely it was fuelled by the hope of a fringe discipline for 'the movement of geography to center stage' (Curry, 1993). Today geography already exploits its next revolution, holography in four-dimensional hyperspace, hailed by an elderly Lord as 'the greatest revolution in geography since the invention of the globe'.

The end of the GIS boom in the late 1990s coincided with a major change of values. What had been a minority opinion in the early 1990s, now became a broad movement: that the most advanced countries in the world could not continue to pursue economic growth forever, but needed to move towards qualitative growth in terms of equity and sustainability (Masser *et al.*, 1992). Political landslides in major European countries brought back the welfare state but also a revival of grassroots democracy.

These developments changed the role of information and by that of geographic information systems in society. People rediscovered that the most important types of knowledge are *not* data and are *not* spatial, but are informal, personal and political, i.e. everything information systems, and GIS in particular, are unable to offer. Some even claimed that the hypothesis that with more and better information all problems could be solved, was itself an expression of a technocratic and functional view of the world (Postman, 1991). All sorts of computerized information systems became associated with everything that was negative: central power, technocracy, the corporate state, police surveillance and organized crime. A wave of violence against computer centres and agencies dealing with data of all sorts swept across Europe. In November 1997 a small group of Luddites set fire to the Eurostat complex in Luxembourg. The fire lasted 5 days, and the smoke trails it generated were recorded by satellites.

Violence cannot solve social conflicts, but in this case it forced the information authorities to radically decentralize and democratize their operations. All cross-links between secret service, police and all sorts of public and corporate data banks were interrupted and put under strict public control. New freedom-of-information acts in many European countries determined that information collected in the public domain had to be made available to the public at no or marginal cost, except where privacy constraints precluded it.

Paradoxically only on first sight, the anti-GIS movement benefited local government GIS. As local self-governance and local planning re-emerged as a central forum of political debate, local government GIS became even more important decision support systems for local land use, transport and environmental planning. In particular the need to redirect urban development towards sustainability gave an unexpected boost to local government GIS as it became apparent that environmental analysis in fields such as air pollution, noise propagation, vegetation, wildlife or micro climate required a more disaggregate spatial scale than conventional aggregate methods.

However, the relationship of local government GIS to power changed. Whereas they were originally designed for the use of the authorities, they now became a public good explicitly designed for public use in an open and participatory process of social experimentation and grassroots decision making. This, of course, required a different type of GIS, one especially designed to be used by non-experts. Therefore a new generation of GIS designed as 'expert systems for non-experts' emerged. Public libraries and institutions of adult education were given a new responsibility as mediators between non-experts and GIS in order to reduce the information gap between the authorities and the public. The result was a revival of public participation in local decision making, in particular in matters of urban planning, and a surge of self-organized user groups exchanging data bases and analytical techniques (cf. Wegener, 1987).

Some say that local planning has become more difficult as public inquiries are more thorough and hence more time-consuming. It is also true that there have been periods of public disinterest when political apathy seemed the ultimate barrier to participatory planning. Moreover, the democratization of knowledge has not solved, but rather acerbated the problem of how to cope with the flood of largely irrelevant information. There even have been instances of deliberate misinformation in the open information arena, and it must be recognized that without the former comprehensive surveillance police work has become less effective. However, most people agree that these are small problems compared with the gain in civic culture.

CONCLUSIONS

Which of the four scenarios is likely to become reality? One view is that there could be different scenarios for different countries. The benign Beyond-GIS scenario, for instance, might have a chance in the mature democracies of north-west Europe, whereas countries with less developed political checks and balances might be at risk of moving into the directions of the Market or Big-Brother scenarios. An opposite view holds that the global competition will bring convergence rather than polarization between countries. In any case it is likely that the future will contain some facets of each of the scenarios. Low-cost GIS software will be widely available and used like spreadsheets and companies will use GIS to increase their profits as in the Market scenario; government agencies will use GIS to process personal spatial data for their purposes as in the Big-Brother scenario and local planning will be changed by access to spatial planning information for everybody as in the Beyond-GIS scenario. Each country can choose to which degree each scenario will come true.

What can be done to enhance the benefits and minimize the dangers of the GIS revolution? The first and most important task is to promote computer literacy and mature and responsible use of GIS through information and education for social consciousness. Like all strategies built on the principle of the enlightened and competent citizen, this may not sound very convincing vis-à-vis powerful economic interests not constrained by moral principles. Therefore good legislation in the area of information is essential. Even though many European countries have made substantial progress towards effective privacy protection, all are sadly lacking in legislation guarding the right of citizens to have access to information collected in the public domain. Recent tendencies to force public agencies to recover the cost of data collection from their users seem to be steps in the wrong direction. Lastly it remains to be seen whether the forthcoming harmonization of privacy and freedom-of-information legislation within the European Union will settle for the lowest common denominator or will bring genuine progress.

These political considerations should, however, not distract from the more fundamental philosophical questions concerning GIS. These questions have hardly found an answer. For instance, it needs to be asked in how far the data model of GIS implies a certain perception of the world and, if applied, will impose that perception on its users. It has been said that because of their US origin many existing GIS not only require their users to communicate in English but also reflect American cultural values (Campari and Frank, 1993; Wegener and Junius, 1993). Lake (1993) claims that the relationship between spatial units of reference and attributes in GIS is essentially positivist, and Curry (1993) points out that current GIS embody the principles of a property-based society. If this is true, GIS would secretly have a conservative and system-stabilizing effect—the direct opposite to their desired innovative and emancipatory role in planning. Under this perspective, the ESRI slogan 'geography organizing our world' takes on an insidious double meaning.

ACKNOWLEDGEMENTS

We are grateful for comments on draft versions of this paper received by several colleagues. In particular Renée E. Sieber, Thanasis Hadzilacos and Francis Harvey provided us with useful suggestions for sharpening our argument and further reading.

REFERENCES

CAMPARI, I. and FRANK, A.U., 1993. Cultural differences in GIS: a basic approach, in Harts, J., Ottens, H.F.L. and Scholten, H.J. (Eds), *EGIS '93 Conference Proceedings*, Vol I, pp. 10–16, Utrecht/Amsterdam: EGIS Foundation.

COUCLELIS, H., 1993. The last frontier, *Environment and Planning B: Planning and Design* **20**, 1–4.

CURRY, M.R., 1993. Producing a new structure of geographical practice: on the unintended impact of geographic information systems, Mimeo, Los Angeles: Department of Geography, University of California at Los Angeles.

HÄGERSTRAND, T., 1967. *Innovation Diffusion as Spatial Process*, Chicago: Chicago University Press.

LAKE, R.W., 1993. Planning and applied geography: positivism, ethics, and geographic information systems, *Progress in Human Geography*, **17**(3), 404–13.

MASSER, I. and ONSRUD, H.J. (Eds), 1993. *Diffusion and Use of Geographic Information Technologies*, Dordrecht: Kluwer.

MASSER, I., SVIDÉN, O. and WEGENER, M., 1992. *The Geography of Europe's Futures*, London: Belhaven Press.

OBERMEYER, N.J., 1992. GIS in democratic society: opportunities and problems, Mimeo. Terre Haute, Indiana: Department of Geography and Geology, Indiana State University.

ONSRUD, H.J., JOHNSON, J.P. and LOPEZ, X., 1994. Protecting personal privacy in using geographic information systems, *Photographic Engineering and Remote Sensing*, **60**, 1083–95.

OPENSHAW, S., 1991. A view on the GIS crisis in geography, or, using GIS to put Humpty-Dumpty back together again, *Environment and Planning A*, **23**, 621–8.

OPENSHAW, S., 1992. Further thoughts on geography and GIS: a reply, *Environment and Planning A*, **24**, 463–6.

OPENSHAW, S., 1993. GIS 'crime' and GIS 'criminality', *Environment and Planning A*, **25**, 451–600.

PICKLES, J. (Ed.) 1994. *Ground Truth: the Social Implications of Geographic Information Systems*, New York: Guildford Press.

POSTMAN, N., 1991. *Technopoly, The Surrender of Culture to Technology*, New York: Alfred Knopf.

ROGERS, E.M., 1962. *The Diffusion of Innovations*, New York: The Free Press.

ROGERS, E.M., 1993. The diffusion of innovations model, in Masser, I. and Onsrud, H.J. (Eds) *Diffusion and Use of Geographic Information Technologies*, pp. 9–24, Dordrecht: Kluwer.

SMITH, N., 1992. History and philosophy of geography: real wars, theory wars, *Progress in Human Geography*, **16**, 257–71.

TAYLOR, P.J. and OVERTON M., 1991. Further thoughts on geography and GIS, *Environment and Planning A*, **23**, 1087–232.

WEGENER, M., 1987. Spatial planning in the information age, in Brotchie, J.F., Hall, P. and Newton, P.W. (Eds) *The Spatial Impact of Technological Change*, pp. 375–92, London: Croom Helm.

WEGENER, M. and JUNIUS, H., 1993. 'Universal' GIS versus national land information traditions: software imperialism or endogenous developments? in Masser, I. and Onsrud, H.J. (Eds) *Diffusion and Use of Geographic Information Technologies*, pp. 213–28, Dordrecht: Kluwer.

CHAPTER 3

Theoretical perspectives on the diffusion of GIS technologies

HEATHER CAMPBELL

INTRODUCTION

Technological developments have a tendency to provoke curiosity, with enthusiasts and sceptics each vociferously attempting to win the hearts and minds of potential users. Researchers from a variety of disciplines have sought to describe and analyse the patterns and processes which lead some innovations to diffuse rapidly throughout society and to a much lesser extent consider why other seemingly worthy developments progress little further than the laboratory (Rogers, 1983; Bijker *et al.*, 1987; Goodman *et al.*, 1990b; Innes and Simpson, 1993). Given the current stage of development of GIS technologies it is now an opportune time to examine general theories concerning the processes influencing diffusion in relation to this specific set of innovations. In the countries with the greatest experience of GIS much of the initial hype often associated with a new technology appears to be passing and there are signs of a more critical approach beginning to develop (see for example, Peuquet and Bacastow, 1991; Barrett, 1992a; John and Lopez, 1992; Innes and Simpson, 1993; Lake, 1993; Pickles, 1994; Campbell, 1996; Campbell and Masser, 1995; Masser and Campbell, 1995). Underlying assumptions about the inevitable advance of GIS technologies and more particularly the association with improved living conditions through the greater availability of information and therefore the democratization of decision making have started to be questioned (see Chapter 2). Furthermore, at both national and international scales indications suggest that the pace of GIS diffusion varies considerably between different contexts (Campbell and Craglia, 1992; Campbell and Masser, 1992).

Differences in the extent of GIS diffusion have been related to variations in national policies on data availability and widely differing institutional arrangements (see Chapter 1). However, evidence suggests that there are some significant similarities between organizations which may be located in widely varying institutional contexts. This raises issues about the role of organizational cultures in determining the propensity for potentially worthy developments to become a taken-for-granted technology. The underlying premise of this debate is that the extent of diffusion of an innovation such as GIS is dependent upon the interaction between that technology and either the external or internal context in which it is located. This is subsequently referred to as social interactionism. It is important to acknowledge, however, that the literature on innovation development points to two further explanatory theories. First, technological determinism based on an assumption that there is an inevitability about the diffusion of new technologies and secondly, economic determinism which links the adoption of innovations to economic progress.

It is recognized at the outset that this chapter cannot do justice to the full range of literature and will inevitably raise more questions as it answers. Furthermore, it is accepted that in placing the emphasis of the subsequent discussion on adoption and therefore regarding this as synonymous with diffusion, creates a false impression. The act of acquiring a GIS, for example, by no means necessarily implies utilization, while the process of implementation, which is crucial in securing the support of users, generally starts sometime before the technology arrives in the organization (Eason, 1988; Campbell 1993). Despite these caveats a simplified definition of diffusion enables clarity to be achieved without undermining the argument. The current trend towards consolidation in the GIS field makes this an appropriate point at which to examine the experiences of GIS diffusion in relation to the existing innovations literature. The concept of the diffusion of innovations has arguably proved of interest to two groups of researchers. First, geographers have shown interest in the spatial aspects of diffusion. Their principal concern has been to investigate the effect of distance from the source of an innovation on the speed and extent of adoption (see, for example, Hägerstrand, 1952, 1967). However, it is researchers based largely within the sociological discipline who have focused most explicitly on examining the underlying processes which account for the diffusion of innovations. It is therefore the findings of this work which is drawn on in the subsequent discussion.

With these considerations in mind the structure of the chapter is as follows. The next section briefly considers the nature of technology and innovations and considers their relationship to GIS. This is followed by a review of the three contrasting perspectives on the key factors influencing the diffusion of technological innovations. These approaches are evaluated with considerable emphasis placed on examining the merits of the social interactionist perspective and in particular the role of organizational cultures in facilitating the diffusion of GIS. This discussion draws on the findings of research conducted in Great Britain into the impact of GIS on local government.

THE NATURE OF TECHNOLOGY

Technology is a notoriously difficult concept to define. There is a tendency for us all to feel we can identify a technology when we see one, although there is no necessary overlap between these perceptions. This lack of consensus is exemplified by the variety of definitions adopted by the contributors to Goodman et al.'s (1990b) book examining the relationship between technology and organizations. However, some common elements can be identified, most particularly that any technology includes knowledge, machines and methods. As a result technology is defined as, 'knowledge of cause-and-effect relationships embedded in machines and methods' (Sproull and Goodman, 1990, p. 255). An important feature of this definition is that it goes beyond the notion of technology as simply items of equipment by placing an emphasis on knowledge or perhaps more particularly an understanding of the role and value of a set of machines and methods. This understanding may be the result of direct experience or equally importantly accepted profession or folk wisdom. Similarly,

Danziger *et al.* (1982) in a study of the implementation of computer based systems in local government in the United States regard technology as a 'package' including hardware and software, people and techniques. The latter has many similarities to Sproull and Goodman's notion of knowledge, as it refers to such features as accepted practices and corporate expectations.

Implicit within much of the discussion about technology is the idea that it is new or innovative. In reality it is generally the machines or methods which are innovative, while the way in which the technology is conceptualized, understood and utilized is based on existing knowledge and practices. Weick (1990) makes a useful distinction between technology and more particularly machines which may be fundamentally new, and technical systems into which this technology is to be embedded which tend not to be. A further implicit assumption about technology is the link with success. Pinch and Bijker (1987) note the findings of Staudenmaier's survey which analysed the content of 25 volumes of *Technology and Culture* and found only nine articles dedicated to the study of failed innovations. As a result there has been a tendency to view the diffusion of technology as a linear and in many ways inevitable process. General experience, however, suggests this to be both misleading and to distort expectations (Rogers, 1986).

Implications for Geographic Information Systems

Implicit within the discussion has been the assumption that GIS should be regarded as a form of technology and perhaps more particularly an innovative technology. There would seem to be considerable support for this notion as GIS combine the key elements of machines, methods and knowledge. In terms of a more detailed categorization, GIS form part of the specialist group of programmable technologies. These technologies which are based on computers, are distinguished from mechanical technologies by their ability to be continuously redesigned (Sproull and Goodman, 1990). Overall, the innovative character of GIS technology would largely seem to be a reflection of developments in computing capabilities. Computer based versions of GIS have been available since the 1960s and in a manual form for many decades and perhaps even centuries before this. However, it is the increased capacity and resulting speed with which computer processors can handle the vast data sets associated with geographical information which has made GIS a commercially attractive product.

It is not uncommon for developments in a related field to prompt innovation in an area which had remained dormant for some time. However, in considering the innovative character of GIS technology the focus must be placed on the machine component as the underlying methods such as overlay and buffering are by no means new nor is the existing knowledge through which individuals make sense of the technology. This understanding of the nature of technology and therefore GIS in turn has important implications for the process of diffusion and the types of factors which are likely to facilitate or inhibit adoption.

THEORIES OF TECHNOLOGICAL DIFFUSION

Consideration of the reasons and impulses as to why societies change, develop and it is hoped, progress has long been of interest. It is implicit within such discussions that progress is associated with the diffusion and widespread acceptance of new practices and technologies. Based on a review of this literature there appear to be three groups of explanatory theories, namely technological determinism, economic determinism and social interactionism. These distinctions inevitably represent a simplification of the rich and varied arguments, however, the aim of the chapter is not to present a detailed review of the literature, rather to identify the key themes and examine these in relation to specific experiences associated with the diffusion of GIS technologies in a European context. The following discussion will examine the arguments concerning technological and economic determinism and then consider these in relation to existing empirical research. The merits of social interactionism will then be considered.

Technological Determinism

Technological determinism assumes that due to the advances inherent within innovations they will diffuse. Put more simply, if someone develops a better washing machine it is bound to sell. Under these assumptions the technical specialists design the new product, users utilize it and the whole of society benefits. As a result the key determinant as to the speed and extent of diffusion are the technical characteristics of the innovation. These assumptions lead to a sense of technological utopianism (see for instance, Naisbitt, 1984; Feigenbaum and McCorduck, 1991). In effect all that is required to create a harmonious and prosperous society are developments in technology. Innovations enable old tasks to be undertaken more effectively as well as opening up new areas of activity. Given the self-evident improvements such facilities offer it is seen as largely inevitable that diffusion will take place; a process which itself results in widespread benefits to both the individuals involved and society as a whole. Such thinking is exemplified by Giuliano's (1991) evaluation of office work in the information age. He envisages office activities as being transformed by computers so as to become flexible and efficient as well as cooperative and interesting. Technology is therefore seen simultaneously as a determinant of change and a force for good.

Given this analysis of technology there are only two real constraints on the widespread diffusion of innovations and in particular computer based systems. The first of these are the technical characteristics of the innovation. Consequently, the lack of adoption of a new technology is regarded as the result of its technical inadequacies such as being too slow or cumbersome. The second set of factors accounting for the non-adoption of a technology is the stupidity or lack of skill of potential users. Technological determinism therefore suggests as Kling (1991) puts it that, 'Perverse or undisciplined people are the main barriers to social reform through computing' (p. 355).

The strengths and weaknesses of this perspective in relation to GIS will be examined later, but it is useful at this stage to highlight the extent to which such thinking pervades the GIS literature. Innes and Simpson (1993) in a survey of articles and abstracts concerning GIS in the *Proceedings of the Urban and Regional Information Systems Association Conference* in the United States point to the nearly all pervasive view that the design of more advanced and necessarily better GIS technologies will inevitably result in adoption. They state:

Often the articles read as if developing more powerful and user-friendly applications will automatically result in the blossoming of GIS in practice

(Innes and Simpson, 1993, p. 230).

Such an emphasis is reflected in the general literature on GIS, the majority of which stresses the great potential of such systems solely based on their technical capabilities. Given the inherent advance over existing methods it is assumed widespread diffusion will occur. Proponents of this line of argument would therefore expect there to be few differences in the level of adoption of GIS between countries in Europe, with any variations resolved by the provision of technical training, for instance.

The growth particularly of the green movement has in recent years provoked a challenge to the utopian claims which some have associated with technology. Schumacher's (1973) work, most especially, highlighted the dark side inherent within what is often referred to as technological progress. So called technological advances such as those linked to the idea of the 'Green Revolution' have been shown to create as many problems as they resolve. As Eckersley puts it,

> ... ecological insights have challenged the technological optimism of modern society and the confident belief that, in time, we can successfully manage all our large scale interventions in natural systems without any negative consequences for ourselves
>
> (Eckersley, 1992, p. 37).

It is paradoxical in this connection that Ryan and Gross' (1943) landmark diffusion study with its inherent pro-innovation bias examined the adoption of hybrid seed corn by farmers in two Iowa communities.

Caution with respect to the potential of technology to improve the human condition is not simply a matter of concern for those involved in the environmental movement. Postman's (1992) image of the transformation of society into 'technopoly' as a result of the proliferation and unthinking use of particular computing technologies vividly illustrates such concerns. Postman stresses that the appropriate application of such technologies is a matter of morality rather than scientific endeavour. Similarly, Roszak (1994) argues that what are required to resolve the complex problems confronting societies are creative ideas rather than machines which simply generate a profusion of information. He suggests, 'Every mature technology brings a minimal immediate gain followed by enormous long-term liabilities' (Roszak, 1994, p. xlvi). In such circumstances it is important to heed the lesson '... that there will never be a machine that leaves us wiser or better or freer than our own naked mind can make us—nor any that helps us work out our salvation with diligence' (Roszak, 1994, p. xlvi).

Economic Determinism

The assumptions discussed above are in many ways implicit within the economic determinist position. What distinguishes this perspective is the emphasis on computerization as the essential next step in economic development. This understanding of the imperative underlying diffusion draws heavily on the work of Toffler (1980) on the third wave. Toffler argues that major transformations in the organization of society are driven by technological changes. So for instance, the 'second wave' was characterized by the change from horse-power to steam-power, and therefore from agricultural to industrial societies, while the impetus for the 'third wave' comes from the widespread diffusion of computers. Computerization, or informatization as it is sometimes referred to, is seen as an inevitable stage in economic growth and thereby the future development of society. Turning the argument round the other way, continued economic growth is an essential part of capitalism and as such is dependent upon technological innovation, the current foundation for a more prosperous society being information technology; as Toffler (1980) himself states:

> What is inescapably clear ... is that we are altering our info-sphere fundamentally ... we are adding a whole new stratum of communication to the social system. The emerging Third Wave info-sphere makes the Second Wave era—dominated by its mass media, the post office, and the telephone—seem hopelessly primitive by contrast   (Toffler, 1980, p. 172).

The need to innovate and by implication computerize is not according to the economic determinist position an option, it is an essential pre-requisite to survival both in the private and public sectors and also at the scale of the individual company or unit of public administration and the nation state. Schumpeter (1950) argued that the need for technological innovation was such a necessity that only companies which could make such activities a routine part of their structure would be able to survive. As a result, large companies were seen as the engines of technological progress and therefore economic growth. In a further development of this general line of thinking Chandler (1977) suggests that the competitive performance of a company is influenced by its internal structure which in turn is determined by the technologies employed for production and communication. [For a more detailed analysis of the argument see Mowery (1990).] Based on these analyses the role of technology and the close association to economic prosperity can be broadened out and applied at the scale of the nation state. Consequently, whether it is argued that countries progress through a number of stages of development linked to the sophistication of the technology employed (see Harbison and Myers, 1959) or that a more random process takes place, the core assumption remains that if a particular nation state is to compete and survive, essentially western production practices must be adopted. As Negandhi is reported by Lynn (1990) as stating, '... the logic of technology is taking over man's differing beliefs and value orientations' (Negandhi in Lynn, 1990, p. 187).

The technical imperative discussed above is developed a stage further by economic determinism into a set of arguments linking technology to economic progress. The diffusion of worthy technologies and in particular computer based systems is seen as a necessary and even inevitable part of the process towards continued economic growth. Similarly the main constraints on the speed of adoption of innovations are their inherent technical capabilities and the shrewdness and intelligence of potential users. These assumptions are again implicit in much of the literature concerning GIS. The benefits of adopting such systems are frequently couched in terms of the greater efficiency offered or the likely cost, staff or time savings that will result from adoption. At a European scale efforts to facilitate technology transfer from the north and west to the south and east reflect arguments which link such activities to accelerated economic growth. In relation to GIS the extremely high proportion of the total number of projects funded by the European Union (EU) in southern Europe exemplifies such trends in thinking. (See for example, Assimakopoulos' findings on Greece in Chapter 9.)

The discussion of economic determinism has so far presented the results of the diffusion of technological innovations, especially computers, in a largely positive light.

It is assumed that technological advance in production and communication will increase prosperity and even the quality of work experiences. Consequently, the general well-being of the whole of society will be enhanced. However, while accepting the underlying rationale that the adoption of new forms of technology is an inherent part of the capitalist mode of production there is a separate strand of argument which sees the results of this process as more sinister and considerably less beneficial to the majority of society. This view suggests that the introduction of new forms of technology will lead to the progressive subordination of the working class as jobs are simplified and made more tedious, whilst surveillance and control over individual employees increases (see for instance, Braverman, 1974; Noble, 1984; Howard, 1985). Such conditions are not limited to the manufacturing sector rather concern about the growing capacity of 'Big Brother' to oversee work is becoming increasingly significant within the traditionally white-collar administrative and clerical sectors due to office automation. Similarly, the growth of computing in all sectors of government has led to concern about reductions in individual freedom due to the increased capacity for personal surveillance (Laudon, 1986). This interpretation of the detrimental impact of the diffusion of innovations is perhaps best illustrated by the anti-utopian view outlined by George Orwell (1949) in his classic text *Nineteen Eighty Four*. The following quote exemplifies the oppressive image of the working environment which such perspectives envisage:

> Behind Winston's back the voice from the telescreen was still babbling ... The telescreen received and transmitted simultaneously. Any sound that Winston made, above the level of a very low whisper, would be picked up by it; moreover, so long as he remained within the field of vision which the metal plate commanded, he could be seen as well as heard ... You had to live ... in the assumption that every sound you made was overheard, and, except in darkness, every movement scrutinized (Orwell, 1949, p. 2).

The underlying rationale of those that express concern about the outcome of the widespread diffusion of technological innovations can be extended to the relations between nation states. At this scale the proposed benefits of technology transfer are re-interpreted into an argument which suggests the promotion of new forms of technology is no more than an attempt by those already dominant within international markets to increase their market share and ensure the dependency of those nations which currently have a lower level of technological capacity. It might be questioned therefore whether by funding GIS projects in Greece or Portugal for instance, the EU is not more profoundly facilitating the production of hardware and software in countries such as Germany, the United States and Great Britain.

The Limitations of Technological Determinism and Economic Determinism

There is always an intrinsic elegance about determinist explanations. By their very nature they construct neat, all consuming models of the future development of society. Furthermore, by concentrating on a single variable the resulting clarity and simplicity of the argument is inherently attractive. However, this essentially one dimensional view inevitably leads to accusations of over-simplification, while the normative character of the underlying argument leaves the explanatory power of the theory open to questioning through empirical research. The proponents of determinist arguments have been criticized for not exposing their theories to the real world situations in which innovations must become embedded if they are to diffuse (Dunlop and Kling, 1991b).

The key premise of the determinist positions outlined above is that worthy technologies will diffuse either because of their inherent technological superiority or due to the imperative of economic growth whether this results in prosperity for the mass of society or for just a few. Implicit within this is the idea that technology has a transforming capacity whether that is of organizational structures and practices at the micro scale or nation states and society at the macro-level. The outcome of the adoption of an innovation is therefore universal whether this be positive as much of the existing literature on GIS suggests or negative as more critical interpretations suggest. The process of diffusion itself is seen as linear in character with for instance utilization naturally progressing from adoption. There is also a tendency to view technology in simplistic terms as just items of machinery and perhaps a set of underlying methods, but the important component of knowledge, which was identified in the definition set out at the start of the chapter, is generally absent. The limitations of these propositions are highlighted by evaluations of empirical research investigating the diffusion of computer based systems in general. Given that GIS exhibit essentially the same characteristics as any other form of information technology (Masser and Onsrud, 1993) these observations provide useful insights into the value of the determinist explanations.

The most striking aspect of reviewing the findings of studies which have examined the diffusion and use of computer based systems in practice, is how messy and complex experiences seem to be. The clarity of the essentially black and white perspectives presented by the determinist positions immediately start to fade into murky grey. Furthermore, the insight gained by exposing existing theories to empirical investigation often appear contradictory and confused. For instance, studies which have examined the ability of computers to transform organizational practices have found that they both centralize and decentralize authority, enrich and routinize jobs and also simultaneously create and destroy employment. [For summaries of these investigations see Kling (1980), Markus and Robey (1988), Rule and Attewell (1991).]

Despite these seeming contradictions these findings highlight a number of important lessons in relation to determinist theories. First, that the outcome of the diffusion of computer based technologies is by no means universal, whether this be in a wholly positive or negative sense. Secondly, given this conclusion it is not the technology itself which determines the level of adoption or the results of implementation but rather the particular circumstances and institutional contexts in which the system is located. As a result while technology may be an instrument of change it does not itself cause that change or even in the vast majority of cases represent the catalyst of change (Danziger *et al.*, 1982; Kraemer and King, 1986; Scott, 1990; Kraemer, 1991; Rule and Attewell, 1991). As Dunlop and Kling (1991c) characterize the impact of information technology on society as a whole:

Computerization is often bound up with the symbolic politics of modernism and rationality. Advanced technologies offer the giddy excitement of adventure with the liberating lure of new possibilities that have few associated problems ... Yet social revolutions are based on changes in ways of life, not just changes in equipment

(Dunlop and King, 1991c, p. 7).

This quote also suggests a third limitation of determinist explanations, that is their virtually exclusive focus on equipment and machinery. Little or no consideration is given to issues concerning existing knowledge or expectations. As a result, especially the more optimistic theories, tend to divorce themselves from the jigsaw of conflict and cooperation into which technological innovations are expected to diffuse and thrive (March and Sproull, 1990). Pinch and Bijker highlight the extent to which different social groupings have very different conceptualizations of the same technology. They suggest '... the socio cultural and political situation of a social group shapes its norms and values, which in turn influence the meaning given to an artifact' (Pinch and Bijker, 1987, p. 46). Consequently, perceptions of the value of a particular innovation are likely to vary greatly even between individuals within the same organization let alone different nations. Such analyses cast further doubt on the generalizability of the determinist explanations. It also suggests the process of diffusion to be far more complicated than the linear view which they present. For instance a technology may diffuse quickly amongst one professional grouping within a particular cultural context and yet fail to be adopted by others which seem to have similar needs. Historical evaluations of technological diffusion therefore indicate the multidirectional character of the process with in many cases the widespread adoption of an innovation dependent not on the quality of the technology itself but rather on social attitudes or institutional developments (Bijker *et al.*, 1987, see in particular Pinch and Bijker's discussion of the diffusion of the bicycle).

As a result despite the inherent elegance and attractiveness of the arguments underlying the technological and economic determinist positions, their explanatory power appears problematic in the face of real world experiences. In some senses these arguments suggest what the trajectory of technological diffusion would be if the world were not as it is. In contrast the social interactionist approach explicitly seeks to examine issues concerning information technology in relation to the social and cultural contexts in which the systems are embedded. The subsequent section will, therefore, examine the nature of this the third of the theoretical positions identified and consider the implications of this approach with respect to the findings of research on GIS in British local government.

SOCIAL INTERACTIONISM

Fundamental to the social interactionist perspective is the assumption that technology is socially constructed. The diffusion of innovations is therefore the result of interaction between the technology and potential users within particular cultural and organizational contexts. [See for example, Mumford and Pettigrew (1975), Markus (1984), Hirschheim (1985), Bijker *et al.* (1987), Hirschheim *et al.* (1987), Eason (1988), Markus and Robey (1988), Klein and Hirschheim (1989), Goodman *et al.* (1990a), Dunlop and Kling (1991c), Innes and Simpson (1993).] Given the huge differences between contexts and widely varying values and motivations of the individuals present within these environments the social interactionist approach suggests the process of diffusion to be complex and problematic. No universal claims are made about the likelihood of the widespread adoption of a computer based technology such as GIS, nor whether the outcome will be beneficial or otherwise to society. Moreover, the outcome is itself regarded as resulting from the socially constructed values underlying adoption rather than the technical characteristics of the innovation. Consequently, even in relation to exactly the same technology the response of potential users is likely to vary considerably.

Implicit within these comments is the notion of technology as combining all three of the elements outlined in the initial definition at the start of the chapter, namely machines, methods and most importantly knowledge. Knowledge itself is not seen as an independent variable but rather to reflect an individual's interpretation of the norms and values of a particular social and/or professional grouping within the context of their society. As a result Sproull and Goodman (1990) suggest that '... technology is a socially constructed reality' (p. 259). Constant (1987) goes further arguing that potential users purchase an image of the technology, particularly the output of that system, rather than the actual items of machinery. In some senses therefore, for a technology to become widely diffused it is more important that it becomes fashionable than that objective measures of its utility indicate a considerable advance over existing techniques. Furthermore, it is likely that a variety of perceptions of the same technology will develop. This view of technology as socially constructed reality also suggests that innovations are not value neutral. The introduction of new computer technology is therefore loaded with social and perhaps even political meaning. For the social interactionist approach questions of ethics and values are not external to the technology but inherent within the system (Ladd, 1991; Mouritsen and Bjorn-Andersen, 1991). Consequently, the diffusion of innovations is dependent upon the interaction between the technology and setting in which it is located. The outcome of this process is again a reflection of contextual imperatives which in turn inevitably influence the style of implementation adopted.

These assumptions about the nature of technology have significant implications for GIS. It is evident even within the academic community that there are considerable differences in accepted technical definitions of what constitutes a GIS. This ambiguity is intensified further when potential or actual users are included. For these individuals GIS tend not to be regarded as a specific collection of computer functionality rather they are seen in terms of applications, output and importantly caricatures of other people's experiences. Moreover, once the technology has been adopted the process of implementation demonstrates the capacity of organizations and individuals to re-invent the technology they have purchased. Re-invention, that is to say the propensity for the same technology to be repeatedly invented in a variety of forms in different organizational settings, is increasingly being seen as a significant part of the diffusion process (Rogers, 1993).

Perceptions of GIS like any other form of technology are highly heterogeneous and it is the underlying understanding implicit within each of these views of the system which influences the extent and speed of diffusion (Goodman et al., 1990b). Rule and Attewell (1991) suggest that computer based applications take one of three forms. These are systems which store information and reproduce it in the same form, secondly those involving some basic arithmetic operations such as inventory control systems and finally the higher level strategic systems which have decision rules built into their operations. This classification highlights the complexity in dealing with GIS, as inherent within much of the software sold under this label are all three types of application. However, while the name may be the same the social construction of the technology in terms of the acceptance of the methods and changes in practices associated with different application fields, are very different. The adoption of a GIS as a decision support system implies a much more profound change to norms and values than as an automated mapping system. Given the importance of the relationship between the technology and the potential contextual setting Innes and Simpson (1993) have cast doubt on whether the current social norms within the planning profession for example, are such as to accept GIS at least in its higher level forms. They applied the characteristics of successful technological innovations as identified by Rogers (1983) of simplicity, observable benefits, relative advantage, ability to make small trials and compatibility, to GIS and found a very poor fit. Regardless of the consequences for GIS, it is noticeable that the features highlighted by Rogers emphasized the importance of the socio-political characteristics of innovations rather than their technical advances.

The discussion has so far concentrated on the nature of technology and the implications of this conceptualization for GIS. However, throughout considerable stress has been placed on the extent to which interaction between technology and the contextual environment influences diffusion. The next section will focus on the role and impact of the context into which innovations such as GIS are to be located.

Context and Diffusion

A key assumption of the social interactionist perspective is that context influences both the diffusion and likelihood of successfully implementing any form of technology. The extent of the emphasis placed on contextual issues varies considerably from those perspectives which simply view environmental considerations as constraints or facilitators of diffusion to those which conceptualize technology as being recreated through the image of the social system in which the particular innovation is embedded. However, regardless of the explanatory power associated with the social context of computing it is first important to examine the nature and scale of the contextual issues included under this heading.

A number of words tend to be used interchangeably with respect to the largely ambiguous concept of context, most particularly environment, setting and culture. Encapsulated within these terms are a vast array of considerations. These vary from the tangible such as level of financial resources available, skills of the workforce and institutional arrangements of central and local government to the more abstract, for instance the values and motivations of individuals, the scope of ethical responsibility and attitudes towards privacy and confidentiality. Perhaps at its most simple, culture is as Deal and Kennedy (1982) describe 'the way we do things around here'.

The scale at which these attributes are considered adds a further dimension to the notion of context. For instance, the key level of analysis extends from the individual through the following increasingly large groupings: a small working unit within a larger company; an organization such as a local authority or private corporation; a professional association; a nation state or even agglomerations of countries such as the Western World or the Pacific Rim. With the exception of the professional association, each has clear institutional boundaries. Professional associations are worth distinguishing separately, as groupings such as surgeons or surveyors, tend to have more in common with fellow professionals than those within the same organizational context. Klein and Hirschheim (1989) note that in addition to variations in what might be described as spatial scale it is also important to bear in mind temporal differences. For instance it is possible that variations in emphasis between nations may simply reflect circumstances at one point in time rather than more profound differences in national attitudes.

There is by no means consensus within studies which adopt what in its widest sense may be termed a socio-technical perspective on technological diffusion, as to the most appropriate emphasis to place on the role of context or the scale at which to focus. With respect to the degree of emphasis placed on culture, Frissen (1989) in a useful review of the literature pertaining essentially to the organizational level of analysis, distinguishes four approaches as follows:

(1) *Culture as a contingency factor.* Under this conceptualization a systematic relationship is envisaged between different variables whereby a variation in any one element has implications for the rest. [See for instance, Child (1985), Hofstede (1980).]

(2) *Culture as a sub-system.* This view suggests culture to exist as a separate sub-system which can be distinguished from others such as work process or management structure. Issues included within the cultural sub-system include, risks, rituals, values and norms, symbols and leadership. [See for instance, Deal and Kennedy (1982), Peters and Waterman (1982).]

(3) *Culture as an aspect-system.* All sub-systems under this approach are viewed as having a cultural dimension. For instance, managerial structure is not just a formal assignment of roles but embodies the values and norms associated with these tasks. [See for instance, March and Olsen (1983,) Meyer and Rowan (1977).]

(4) *Organization as a cultural phenomenon.* The central proposition of this view is that an organization for example, 'does not have a culture, but it is a culture' (Frissen, 1989, p. 572). Culture is seen as the very essence of an organization rather than one, although perhaps a highly significant, aspect. It is, therefore, impossible to separate an organization from its cultural context as they are one and the same.

It is implicit within these descriptions that the nature of the contextual considerations on which each focuses becomes less tangible in character as the focus moves from culture as simply a contingency factor to the organizational unit of analysis to itself as a cultural phenomenon.

In terms of GIS diffusion the literature is as yet limited. However, in the studies which have been conducted culture is generally treated as a contingency factor (Onsrud and Pinto, 1991; Craglia, 1993). Furthermore, culture tends to be viewed simply in terms of national characteristics, which in turn account in some measure for the extent and speed of GIS diffusion. As a result stress is placed on the tangible products of a particular culture such as the institutional framework rather than the nature of the culture itself (see the discussions later in this book). It is also unclear in much of this work what precisely is the unit of analysis. Regardless of whether culture is seen as an all pervasive phenomenon or as just one amongst a number of variables, it is important to distinguish the level of focus. For instance, diffusion research in general has focused on individuals as the key unit of study. In the case of GIS some doubt has been cast as to the appropriateness of such an approach, for it is rarely a single individual who decides whether to adopt such systems (Campbell, 1991; Rogers, 1993). The decision processes which determine whether it is appropriate, or equally inappropriate, to acquire a GIS are embedded within organizations.

In practical terms therefore there is a justification for focusing on organizations as the unit of study for research into GIS diffusion. There is also theoretical support for such an approach. Organizations come at the intersection between the various levels. They act as the arena in which the culturally influenced or some might say determined values and norms of individuals and professional groupings are played out. Sproull and Goodman (1990) argue that it is at the intersection between levels that the greatest insight can be gained. Furthermore, there is considerable evidence to suggest that through the interaction of elements within this context organizations themselves develop their own culture (Deal and Kennedy, 1982; Kanter, 1988; Handy, 1991, 1993).

> Therefore, understanding systems development and use involves understanding the relationships between organizational participants and groups. Very importantly, this involves understanding the micro-politics of any social system that is reproduced by differentially motivated agents and the broader social contradictions that will contextualize the forms of agreements that can be achieved
>
> (Mouritsen and Bjorn-Andersen, 1991, p. 312).

The remainder of the chapter will therefore, seek to identify the key aspects of organizational cultures for GIS diffusion and consider the appropriateness and insight provided by such an approach.

Organizational Cultures

The preceding argument suggests that organizations are the key unit of analysis. The widespread diffusion of GIS is dependent upon the acceptance of the merits of the technology within each organizational arena. It is therefore within these contexts that innovations become socially constructed and decisions are made about their adoption, implementation and utilization. For the determinist explanations outlined at the start of the chapter, organizations are of little concern as they are all seen as essentially the same. In terms of the metaphors developed by Morgan (1986) organizations tend to be viewed as either machines or organisms by those adopting an essentially rational view and as coercive political entities by the more critical interpretations. In each case individual organizations are regarded as largely passive. Morgan, however, distinguishes five other metaphors, typified by the notion of organizations as cultures. Under this conceptualization the characteristics of organizations are not regarded as universal. As Barrett (1992b) points out organizations responsible for similar activities within the same legal framework such as those in the public sector or all those involved with the manufacture of cars will have elements in common, while the operational application of this framework varies between organizations within the same jurisdiction. Despite differences in the legislative framework between countries it appears that the evolution of particular organizational cultures is dependent upon how the external constraints are internalized and understood within the organization. It is quite possible therefore that there may be significant similarities between organizations which are subject to very different institutional arrangements. Consequently, organizational culture can be usefully conceptualized as '... the complex of patterns of meaning in and around an organization' (Frissen, 1989, p. 572).

There appear to be two inter-related elements of organizational cultures which are important in relation to the diffusion of GIS. These are styles of bureaucracy and the approach to decision making. These two aspects are closely related with the first referring to the norms and values of the organization which are reflected in, for instance, routine practices, general expectations, styles of leadership and the staffing structure, while the second focuses on the formal and informal procedures of decision making and in particular the role of information in this process. In relation to the propensity for organizations to adopt GIS the key aspect with respect to the style of bureaucracy appears to be the capacity for the organization and by implication the individuals within that environment to cope with change, and with regard to decision making the identification of what constitutes proof in that particular context and whether GIS is seen as contributing to this. Each of these will be examined in turn. [For a full discussion of the nature of organizations and decision making in relation to GIS see Campbell and Masser (1995).]

Uncertainty and instability are inherent to all organizations. However the response to such pressures varies immensely between different contexts. Some demonstrate a capacity to absorb change and even create their own forms of instability by adopting new or innovative practices or policies. In contrast, the vast majority appear to be uneasy in the face of change preferring to maintain existing routines rather than risk the upheaval and uncertainty which will be induced. Much of the work in this area has been undertaken in relation to the private sector, most particularly examining the relationship between the presence of so called innovative cultures and national economic performance (Kanter, 1988). Such studies demonstrate that a propensity for innovation is relatively uncommon even within the private sector.

Organizations are often described as negotiated orders. This highlights the extent to which organizations consist of a heterogeneous set of conflicting values and interests. In such circumstances any change, for example the introduction of a GIS, must be agreed through negotiation with all the varied interests. As Barrett and McMahon (1990) have noted in health service organizations:

> ... shifts in policy or demands for services are not spotted solely from the windows of the Chief Executive's suite of offices! The whole organization is full of strategists, peering through windows at all levels of the organization, all scanning, anticipating, planning and adapting: all seeing how primary changes will either be disadvantageous or offer opportunities for advancement. So the true impact of change 'out there' is not felt at some imaginary organizational boundary, but instead it is felt 'in here' in how it affects the interests, power relationships and bargaining tactics of all the organizations' partisans   (Barrett and McMahon, 1990, p. 262).

As a result for GIS to be adopted by an organization and successfully implemented it must involve either the very minimum of change or the change must be accepted by all the separate interests within the organization.

The second aspect of organizational culture which is important for GIS diffusion is the ability to identify what constitutes proof in a particular environment and in so doing establish the contribution to be made by the products of a GIS. Undoubtedly the nature of proof is likely to change according to the type of activity being considered. For instance, for routine tasks such as responding to simple queries the evidence required to support a particular decision is likely to take a different form than for strategic or policy related issues. It is important in considering this issue to remember that formal information flows along less structured channels of communication (Mouritsen and Bjorn-Andersen, 1991) and that similarly there are often significant differences between espoused theories and the actual theories in use as to the value and merit of conventional data sources (Argyris and Schon, 1974).

Much of the literature which has examined the use of the information produced by computer based systems emphasizes the extent to which it reinforces both existing practices and the power structure of an organization [see for example, Danziger *et al.* (1982), Frissen (1989), Kraemer (1991), and Pfeffer (1981)]. By implication therefore it might be assumed that information systems will diffuse if the main interests within the organization are confident that the information produced will support their activities and values. March and Sproull (1990) note that it is not necessarily the technical quality of the information generated which decision makers are interested in but rather the social status associated with the system. They go on to conclude that the symbolic value of owning a system may be more important than its actual use. This is perhaps most evident in the case of complex strategic decisions. It is in relation to this type of decision that the contribution of GIS has been regarded in much of the literature to be most profound. However, in practical terms decision support systems often lack credibility in the eyes of decision makers as they assume circumstances to be stable over time. Problems change and the information which constituted proof yesterday may be regarded as inadequate today. In many cases the answer to a particular problem is already known and the decision maker is simply in search of a justification. Proof with regard to decision making is very often as much symbolic as real. 'Information in organizations is not innocent; rather it is shaped by the expectations of its consequences' (March and Sproull, 1990, p. 151).

This section has emphasized the importance of organizational culture in relation to the diffusion of technology and in particular drawn attention to two important aspects of organizational cultures. These are an ability to cope with change and identifying what constitutes proof within a particular environment. These issues will now be considered in relation to the findings of research in a number of contexts but particularly British local government.

RESEARCH FINDINGS

During 1991 and 1992 12 detailed case studies were undertaken as part of a project investigating the impact of GIS on British local government. This study represents one of the few systematic attempts to examine the processes influencing the diffusion and effective utilization of GIS in practice. [For a detailed discussion of the research methodology and findings see Campbell (1994), Campbell and Masser (1995).] One of the most striking findings of this research was the very low level of utilization of GIS facilities within the case studies even though the technology had been present within the organization for at least 2 years and in most instances 4 or 5. This suggests that diffusion is not simply a matter of getting the computing technology inside the organization. While this raises important issues for the implementation of GIS technologies the subsequent evaluation concentrates on the processes influencing the adoption decision. In addition the discussion will consider some of the findings of two, admittedly more limited studies which have been undertaken by the author in the United States in 1991 and 1994 (Campbell, 1991) and Portugal.

Evaluation of the Empirical Evidence

The findings of the British case studies exemplify the complex nature of the initial motivations surrounding the decision to purchase a GIS. In two-thirds of the case studies the prime reason for the introduction of a GIS appeared to be social and political rather than technical in nature. In the majority of these cases GIS technology was acquired at the instigation of a new chief executive or head of department. The implication being that the introduction of computer based technologies such as GIS marked a change in approach towards a more progressive and modern administration. In a couple of cases GIS were also seen as asserting some form of independence from a central department such as computing services. The link between GIS and social status therefore appeared crucial. Only in two cases did the formal justification as to the technical benefits of GIS expressed in committee reports appear to be borne out by the detailed investigations. Recent reports on these systems suggest they have encountered some of the most severe difficulties in achieving utilization as the technical advances have proved harder than expected to realize. If these benefits cannot be delivered the technology has very little value to these contexts.

These were not, however, the only issues which appeared to account for the take-up of GIS in the case studies. In two instances there seemed to be a more profound and deep-rooted explanation, namely an inherent capacity, perhaps even desire, to innovate. This cultural imperative led these organizations to be in the vanguard, if not the initiators of new policies and procedures throughout the whole organization. Consequently, the adoption of GIS may be viewed as simply one part of a process which is deeply embedded within the culture of the organization. An important aspect of this keenness to embrace innovation is the acceptance of the problematic and complex nature of the process of introducing a new set of procedures or policy. The British research indicates that the implementation of GIS technologies is a long and uncertain process. It is, therefore, vital that the organization can sustain the uncertainties associated with this process beyond the initial flush of enthusiasm which results from securing the go-ahead to purchase the equipment. Uncertainty and change are inherent within organizational life but are very often regarded as threats rather than opportunities. This tendency leads many organizations to stop or withhold resources once a situation starts to become problematic. In contrast those which might be described as inherently innovative demonstrate a capacity to sustain such activities, learn from their mistakes and move the project forward, rather than strangle and thwart initiative. It is implicit that the majority of individuals within such organizations share this perspective. It appears, therefore, that innovative organizational cultures attract innovative people and consequently there is a tendency to perpetuate and recreate these conditions. Much more research is required into the nature of organizational cultures, but impressionistic evidence from Europe and North America suggests that innovative environments regardless of national boundaries have more in common than they have with organizations in the same country.

The discussion of organizational culture identified the conceptualization of decision making within organizations to have an important influence on the diffusion and development of technologies such as GIS. The findings of the British study and reports of trends in other countries suggest that GIS are being applied to relatively simple tasks such as displaying data or search and query activities. In terms of Rule and Attewell's (1991) classification these represent the least transformative activities, where the resulting decisions are generally uncontroversial and focus on operational rather than managerial or strategic activities. The organizations which have achieved the utilization of GIS have established that it is in these areas that the type of information produced by GIS will be accepted by the members of that organization as proof. In contrast much of the GIS literature focuses on strategic decision making as the area in which the output of GIS can make the greatest contribution. This in turn assumes decision making to be a rational process in which the appropriate course of action is determined on the basis of conventional forms of analysis and information. Evidence from a range of contexts including both the public and private sectors casts doubt on the appropriateness of such a description. [See for instance, Argyris (1971), Greenberger *et al.* (1976), Weiss (1977), Lindblom and Cohen (1979), Feldman and March (1981), Larsen (1985), March (1987), Sabatier and Jenkins-Smith (1988), Campbell (1990).] In relation to GIS the experiences of those few organizations which have attempted to develop applications orientated towards decision support are highly instructive.

British, Portuguese and American examples of GIS applications which have a decision support orientation and are being utilized, all demonstrate the capacity of those developing the systems to identify the symbolic rather than the instrumental role of information in decision making. The information generated by the GIS is not used to determine the appropriate course of action but rather to support the decision which has already been made on the basis of other evidence. For instance, in one case the realization by a senior elected member that the possession of supporting evidence and the display of that material on a map was the key to achieving additional funding from central government resulted in substantial investment in GIS technology. In other words the individual had identified what constituted proof in that context and saw that a GIS could make a contribution. This contrasts with experience in another environment where a less senior politician who was also a newcomer to the area attempted to use the output from a GIS to inform the decision making process within the authority. The result was that the other elected officials had the maps locked away as they were suspicious about their value and lacked confidence in their own ability to understand the information. The output of a GIS in this context did not accord with existing social practices and consequently did not constitute proof. These examples demonstrate the importance of identifying what constitutes proof in a particular context. The lack of use of GIS for strategic decision making suggests that in many cases the output of GIS is regarded as having little contribution to make and where such applications have been developed the information produced is seen as having a symbolic rather than instrumental role.

These findings suggest that the intrinsic qualities of organizations have a significant role in influencing the extent and speed of GIS diffusion. The outcome of this process appears to depend upon the inter-relationship between the technology and the organizational culture. The findings also suggest that diffusion is a multi-stage process in which the initial decision to purchase the equipment is in many ways not the most problematic aspect. Far more uncertain is the diffusion and by implication the acceptance of the technology by all the various sectors and interests within the organization.

CONCLUSIONS

The development of any new technology provokes interest which is often reflected in a questioning of the likely implications of such systems for future work practices and core values of society. The debate surrounding GIS technology tends to mirror this trend with the lofty technical claims tempered by the more cautious views of experiences in practice. It has not been the aim of this paper to predict the future trajectory of GIS developments in Europe. Rather the discussion has attempted to set the current lines of argument in context and to stimulate the debate by presenting a perspective on technological diffusion which appears to be largely overlooked in much of existing GIS literature, namely the social interactionist approach. It is inevitable that such a discussion will raise as many questions as it answers.

A review of the general literature on the diffusion of innovations suggests there to be three explanatory theories as to the reasons why a new technology will diffuse. The technological and economic determinist positions exhibit a clarity of argument which is attractive but loses much of its credibility when exposed to empirical evaluation. In contrast the social interactionist position lacks a certain elegance but seems to offer more in terms of accounting for the varied experiences of individuals and organizations in practice. The precise nature of the relationship between technology and the multi-levelled contexts in which it is located is unclear. Both Constant (1987) and Lynn (1990) for instance, reflecting on the field of innovation research in general, suggest that there is urgent need for further work. As Lynn states:

> In short, we need organizational theory and organizational research that better encompass national and organizational cultures, technology, organizational structure and performance   (Lynn, 1990, p. 195).

The final part of this chapter attempts to inform the process of devising a research agenda with respect to GIS diffusion. This tentative analysis of the experiences of GIS indicates the importance of organizational cultures. Factors such as the institutional framework and the availability of spatial data may influence the speed of GIS diffusion but what is more vital is how these issues are translated and understood within individual organizations. In terms of the key characteristics of organizations attention has been drawn to the issues of innovativeness and the nature of proof within the decision making process. Technologies are not value neutral and an understanding of the implications of this for organizations and societies would undoubtedly inform the debate surrounding GIS.

With these considerations in mind there seems to be a number of fruitful lines of research which would inform understanding of both GIS and the diffusion of technology more generally:

(1) The identification of the key characteristics of organizational culture for the diffusion of GIS.

(2) An evaluation of the nature and impact of organizational culture through an examination of one professional grouping.

(3) Consideration of what makes organizations innovative and whether it is possible to create such cultures.

(4) An assessment of the contribution of GIS to various fields of decision making.

(5) An analysis of perspectives on the nature of GIS technology with particular emphasis on the extent to which each organization re-invents a particular form of technology.

(6) The ethical implications surrounding the adoption, implementation and utilization of GIS technologies.

ACKNOWLEDGEMENTS

I am very grateful to Sue Barrett and Michael Wegener for their comments which greatly assisted in the preparation and refinement of the ideas expressed in this chapter.

REFERENCES

ARGYRIS, C., 1971. Management information systems: the challenge to rationality and emotionality, *Management Science*, **17**(6), B275–B292.

ARGYRIS, C. and SCHON, D. 1974. *Theory in Practice: Increasing Professional Effectiveness*, San Francisco: Jossey-Bass.

ASSIMAKOPOULOS, D., 1993. The Greek GIS community, *Proceedings of the 5th European GIS Conference*, Genoa, 29 March–1 April, pp. 723–32, Utrecht: EGIS Foundation.

BARRETT, S.M., 1992a. Implementing geographic information systems within organisations, paper presented at European Advanced Workshop in Implementing GIS, Padua, May.

BARRETT, S.M., 1992b. Information technology and organisational culture: implementing change, paper presented at EGPA Conference, Maastricht, August.

BARRETT, S. and MCMAHON, L., 1990. Public management in uncertainty: a micro-political perspective of health service in the United Kingdom, *Policy and Politics*, **18**(4), 257–68.

BIJKER, W.E., HUGHES, T.P. and PINCH, T.J. (Eds), 1987. *Social Construction of Technological Systems: New Directions in the Sociology and History of Technology*, Cambridge, MA: MIT Press.

BRAVERMAN, H., 1974. *Labor and Monopoly Capital*, New York: Monthly Review.

CAMPBELL, H.J., 1990. 'The use of geographic information in local authority planning departments', doctoral thesis, Department of Town and Regional Planning, Sheffield: University of Sheffield.

CAMPBELL, H.J., 1991. Organisational issues in managing geographic information, in Masser, I. and Blakemore, M. (Eds) *Handling Geographic Information*, pp. 259–82, London: Longman.

CAMPBELL, H.J., 1992. Organisational issues and the implementation of GIS in Massachusetts and Vermont: some lessons for the UK, *Environment and Planning B*, **19**(1), 85–95.

CAMPBELL, H.J., 1993. Successful GIS implementation: the impossible dream? *Proceedings of the 16th Urban Data Management Symposium*, Vienna, 6–10 September, pp. 144–57, Delft: Urban Data Management Society.

CAMPBELL, H.J., 1994. How effective are GIS in practice? A case study of British local government, *International Journal of Geographical Information Systems*, **8**(3), 309–25.

CAMPBELL, H.J., 1996. A social interactionist perspective on computer implementation, *Journal of the American Planning Association* (in press).

CAMPBELL, H.J. and CRAGLIA, M., 1992. The diffusion and impact of GIS on local government in Europe: the need for a European-wide research agenda, *Proceedings of the 15th Urban Data Management Symposium*, Lyon, 16–20 November, pp. 133–54, Delft: Urban Data Management Society.

CAMPBELL, H.J. and MASSER, I., 1992 GIS in local government: some findings from Great Britain, *International Journal of Geographical Information Systems*, **6**(6), 529–46.

CAMPBELL, H.J. and MASSER, I., 1993. The impact of geographic information systems on British local government, in Mather, P. (Ed.) *Geographic information handling*, Chichester: John Wiley.

CAMPBELL, H.J. and MASSER, I., 1995. *GIS and Organisations*, London: Taylor and Francis.

CHANDLER, A.D., 1977. *The Visible Hand*, Cambridge, MA: Harvard University Press.

CHILD, J., 1985. Managerial strategies, new technology and labour process, in Knights, D., Willmott, H. and Collinson, D. (Eds) *Job Redesign, Critical Perspectives on the Labour Process*, pp. 105–41, Aldershot: Gower.

CONSTANT II, E.W., 1987. The social locus of technological practice: community, system or organisation? in Bijker, W.E., Hughes, T.P. and Pinch, T.J. (Eds) *The Social Construction of Technological Systems: New Directions in the Sociology and History of Technology*, pp. 223–42, Cambridge, MA: MIT Press.

CRAGLIA, M., 1993. 'Geographical information systems in Italian municipalities: a comparative analysis', doctoral thesis, Department of Town and Regional Planning, Sheffield: University of Sheffield.

DANZIGER, J.N., DUTTON, W.H., KLING, R. and KRAEMER, K.L., 1982. *Computers and Politics: High Technology in American Local Government*, New York: Columbia University Press.

DEAL, T.E. and KENNEDY, A.A., 1982. *Corporate Cultures: the Rites and Rituals of Corporate Life*, Reading: Addison-Wesley.

DUNLOP, C. and KLING, R. (Eds) 1991a. *Computerization and Controversy: Value Conflicts and Social Choices*, San Diego: Academic Press.

DUNLOP, C. and KLING, R., 1991b. The dreams of technological utopianism, in Dunlop, D. and Kling, R. (Eds) *Computerization and Controversy: Value Conflicts and Social Choices*, p. 14–30, San Diego: Academic Press.

DUNLOP, C. and KLING, R., 1991c. Social controversies about computerization, in Dunlop, C. and Kling, R. (Eds) *Computerization and Controversy: Value Conflicts and Social Choices*, p. 1–12, San Diego: Academic Press.

EASON, K., 1988. *Information Technology and Organisational Change*, London: Taylor and Francis.

ECKERSLEY, R., 1992. *Environmentalism and Political Theory: Toward an Ecocentric Approach*, London: UCL Press.

FEIGENBAUM, E. and MCCORDUCK, P., 1991. Excerpts from 'The fifth generation: artificial intelligence and Japan's computer challenge to the world', in Dunlop, C. and Kling, R. (Eds) *Computerization and Controversy: Value Conflicts and Social Choices*, p. 31–54, San Diego: Academic Press.

FELDMAN, M.S. and MARCH, J.G., 1981. Information in organisations as signal and symbol, *Administrative Science Quarterly*, **26**, 171–86.

FRISSEN, P.H.A., 1989. The cultural impact of informatization in public administration, *International Review of Administrative Sciences*, **55**, 569–86.

GIULIANO, V., 1991. The mechanisation of office work, in Dunlop, C. and Kling, R. (Eds) *Computerization and Controversy: Value Conflicts and Social Choices*, pp. 200–12, San Diego: Academic Press.

GOODMAN, P.S., GRIFFITH, T.L. FENNER, D.B., 1990a. Understanding technology and the individual in an organisational context, in Goodman, P.S., Sproull, L.S. and Associates (Eds) *Technology and Organisations*, pp. 45–86, San Francisco: Jossey-Bass.

GOODMAN, P.S., SPROULL, L.S. ASSOCIATES, 1990b. *Technology and Organisations*, San Francisco: Jossey-Bass.

GREENBERGER, M., GRENSON, M.A. and CRISSEY, B.L., 1976. *Models in the Policy Process: Public Decision Making in the Computer Era*, New York: Russell Sage Foundation.

HÄGERSTRAND, T., 1952. The propagation of innovation waves, *Lund Studies in Geography*, **4**.

HÄGERSTRAND, T., 1967. *Innovation Diffusion as a Spatial Process*, Chicago: University of Chicago Press.

HANDY, C.B., 1991. *Gods of Management*, London: Souvenir.

HANDY, C.B., 1993. *Understanding Organisations*, Harmondsworth: Penguin.

HARBISON, F. and MYERS, C.A., 1959. *Management in the Industrial World*, New York: McGraw-Hill.

HIRSCHHEIM, R.A., 1985. *Office Automation: a Social and Organisational Perspective*, Chichester: John Wiley.

HIRSCHHEIM, R., KLEIN, H. and NEWMAN, M., 1987. Information system development as social action: theory and practice, RDP 87/6, Oxford, Oxford Institute of Information Management, University of Oxford.

HOFSTEDE, G., 1980. *Culture's Consequences*, Beverley Hills, CA: Sage.

HOWARD, R., 1985. *Brave New Workplace*, New York: Viking Penguin.

INNES, J.E. and SIMPSON, D.M., 1993. Implementing GIS for planning: lessons from the history of technological innovation, *American Planning Association Journal*, **59**(2), 230–6.

JOHN, S.A. and LOPEZ, X.R., 1992. Integrating data derived from within the British local government institutional framework, *Proceedings of the Association for Geographic Information Conference*, Birmingham, 27–29 November, pp. 1.18.1– 1.18.8, Rickmansworth: Westrade Fairs Ltd.

KANTER, R.M., 1988. *The Change Masters: Corporate Entrepreneurs at Work*, London: Routledge.

KLEIN, H.K. and HIRSCHHEIM, R.A., 1989. Legitimation in information systems development: a social change perspective, *Office, Technology and People*, **5**(1), 29–46.

KLING, R., 1980. Social analysis of computing: theoretical perspectives in recent empirical research, *Computing Surveys*, **12**(1), 61–110.

KLING, R., 1991. Computerization and social transformations, *Science, Technology and Human Values*, **16**(3), 342–67.

KRAEMER, K.L., 1991. Strategic computing and administrative reform, in Dunlop, C. and Kling, R. (Eds) *Computerization and Controversy: Value Conflicts and Social Choices*, pp. 167–80, San Diego: Academic Press.

KRAEMER, K.L. and KING, J.L., 1986. Computing and public organisations, *Public Administration Review*, **46**, 488–96.

LADD, J., 1991. Computers and more responsibility: a framework for an ethical analysis, in Dunlop, C. and Kling, R. (Eds) *Computerization and Controversy: Value Conflicts and Social Choices*, pp. 664–75, San Diego: Academic Press.

LAKE, R.W., 1993. Planning and applied geography: positivism, ethics and geographic information systems, *Progress in Human Geography*, **17**(3), 404–13.

LARSEN, J.K., 1985. Effect of time on information utilization, *Knowledge: Creation, Diffusion, Utilization*, **7**(2), 143–59.

LAUDON, K. 1986. *Dossier Society*, New York: Columbia University Press.

LINDBLOM, C.E. and COHEN, D.K., 1979. *Usable Knowledge: Social Science and Social Problem Solving*, New Haven: Yale University Press.

LYNN, L.H., 1990. Technology and organisations: a cross-national analysis, in Goodman, P.S., Sproull, L.S. and Associates (Eds) *Technology and Organisations*, pp. 174–99, San Francisco: Jossey-Bass.

MARCH, J.G., 1987. Ambiguity and accounting: the elusive link between information and decision making, *Accounting, Organisations and Society*, **12**(2), 153–68.

MARCH, J.G. and OLSEN, J.P., 1983. Organizing political life: what administrative reorganisation tells us about government, *The American Political Science Review*, **77**(2), 281–96.

MARCH, J.G. and SPROULL, L.S., 1990. Technology, management and competitive advantage, in Goodman, P.S., Sproull, L.S. and Associates (Eds), *Technology and Organisations*, pp. 144–73, San Francisco: Jossey-Bass.

MARKUS, L.M., 1984. *Systems in Organisations: Bugs and Features*, Boston MA: Pitman.

MARKUS, L. and ROBEY, D 1988. Information technology and organisational change: causal structure in theory and research, *Management Science*, **34**(5), 583– 98.

MASSER, I. and CAMPBELL, H., 1995. Information sharing: the effect of GIS on British local government, in Onsrud, H. and Rushton, G. (Eds) *Institutions Sharing Geographic Information*, pp. 230–49, New Brunswick: Rutgers University Press.

MASSER, I. and ONSRUD, H.J. (Eds) 1993. *Diffusion and the Use of Geographic Information Technologies*, Dordrecht: Kluwer.

MEYER, J.W. and ROWAN, B., 1977. Institutionalized organisations: formal structure as myth and ceremony, *American Journal of Sociology*, **83**(2), 340–63.

MORGAN, G., 1986. *Images of Organisation*, Beverley Hills: Sage.

MOURITSEN, J. and BJORN-ANDERSEN, N., 1991. Understanding third wave information systems, in Dunlop, C. and Kling, R. (Eds) *Computerization and Controversy: Value Conflicts and Social Choices*, pp. 308–20, San Diego: Academic Press.

MOWERY, D.C., 1990. Technology and organisations: an economic/institutional analysis, in Goodman, P.S., Sproull, L.S. and Associates (Eds) *Technology and Organisations*, pp. 200–31, San Francisco: Jossey-Bass.

MUMFORD, E. and PETTIGREW, A., 1975. *Implementing Strategic Decisions*, London: Longman.

NAISBITT, J., 1984. *Megatrends*, London: McDonald.

NOBLE, D.F., 1984. *Forces of Production: a Social History of Industrial Automation*, New York: Knopf.

ONSRUD, H.J. and Pinto, J.K., 1991. Diffusion of geographic information innovations, *International Journal of Geographical Information Systems*, **5**, 447–67.

ORWELL, G., 1949. *Nineteen Eighty Four*, Harmondsworth: Penguin.

PETERS, T.J. and Waterman, R.H., 1982. *In Search of Excellence: Lessons from America's Best-Run Companies*, New York: Harper and Row.

PEUQUET, D.J. and BACASTOW, T., 1991. Organisational issues in the development of geographical information systems: a case study of U.S. Army topographic information automation, *International Journal of Geographic Information Systems*, **5**(3), 303–19.

PFEFFER, G., 1981. *Power in Organisations*, London: Pitman.

PICKLES, J. (Ed.) 1994. *Ground Truth: the Social Implications of Geographical Information Systems*, New York: Guildford Press.

PINCH, T.J. and BIJKER, W.E., 1987. The social construction of facts and artifacts: or how the sociology of science and the sociology of technology might benefit each other, in Bijker, W.E., Hughes, T.P. and Pinch, T.J. (Eds) *The Social Construction of Technological Systems: New Directions in the Sociology and History of Technology*, pp. 17–50, Cambridge: MIT Press.

POSTMAN, N., 1992. *Technopoly: The Surrender of Culture to Technology*, New York: Knopf.

ROGERS, E.M., 1983. *Diffusion of Innovations*, New York: Free Press.

ROGERS, E.M., 1986. *Communication Technology: the New Media in Society*, New York: Free Press.

ROGERS, E.M., 1993. The diffusion of innovations model, in Masser, I. and Onsrud, H.J. (Eds) *Diffusion and Use of Geographic Information Technologies*, pp. 9–24, Dordrecht: Kluwer.

ROSZAK, T., 1994. *The Cult of Information: a Neo-Luddite Treatise of High-Tech, Artificial Intelligence, and the True Art of Thinking*, Berkeley, CA : University of California Press.

RULE, J. and ATTEWELL, P., 1991. What do computers do?, in Dunlop, C. and Kling, R. (Eds) *Computerization and Controversy: Value Conflicts and Social Choices*, pp. 131–49, San Diego: Academic Press.

RYAN, B. and GROSS, N.C., 1943. The diffusion of hybrid seed corn in two Iowa communities, *Rural Sociology*, **8**, 15–24.

SABATIER, P.A. and JENKINS-SMITH, H.C., 1988. Symposium editors' introduction, *Policy Sciences*, **21**, 123–7.

SCHUMACHER, E.F., 1973. *Small is Beautiful*, London: Harper and Row.

SCHUMPETER, J., 1950. *Capitalism, Socialism and Democracy*, New York: Harper and Row.

SCOTT, W.R., 1990. Technology and structure: an organisational-level perspective, in Goodman, P.S., Sproull, L.S. and Associates (Eds) *Technology and Organisations*, pp. 109–43, San Francisco: Jossey-Bass.

SPROULL, L.S. and GOODMAN, P.S., 1990. Technology and organisations: integration and opportunities, in Goodman, P.S., Sproull, L.S. and Associates (Eds) *Technology and Organisations*, pp. 254–65, San Francisco: Jossey-Bass.

TOFFLER, A., 1980. *The Third Wave*, New York: Random House.

WEICK, K.E., 1990. Technology as equivoque: sensemaking in new technologies, in Goodman, P.S., Sproull, L.S. and Associates (Eds), *Technology and Organisations*, pp. 1–44, San Francisco: Jossey-Bass.

WEISS, C.H. (Ed.), 1977. *Using Social Science Research in Public Policy Making*, D.C. Heath and Co: Lexington, MA, Lexington Books.

PART II

GIS in Local Government in the Five European Countries Surveyed Using a Similar Methodology

CHAPTER 4

Great Britain: the dynamics of GIS diffusion

IAN MASSER and HEATHER CAMPBELL

INTRODUCTION

The comprehensive survey of the diffusion of GIS in British local government that was carried out by the authors between February and June 1991 provided a model for the four national surveys whose findings are reported in subsequent chapters [see, for example, Campbell and Masser (1992), Masser and Campbell (1993)]. This survey was undertaken as part of a larger research project investigating the impact of GIS on British local government. The aim of this research is to combine an overview of GIS adoption with in-depth analysis of the factors influencing implementation in specific organizations (Campbell and Masser, 1995).

The present chapter describes and evaluates the findings of a second comprehensive survey of all 514 British local authorities that was carried out for the Local Government Management Board between June and September 1993 using the same methodology as the earlier survey. These findings are broadly comparable in terms of time with those reported for other European countries in subsequent chapters. They are also of considerable interest in their own right because it is possible to make direct comparisons between them and those from the earlier survey. As a result they provide some unique insights into the dynamics of the GIS process in British local government.

The chapter is divided into six main parts. The first two of these describe some of the distinctive features of the structure of local government and topographic data provision in Great Britain respectively. The findings of the 1993 survey with respect to the adoption and take-up of GIS are summarized in the third and fourth parts. The fifth part of the chapter explores the dynamics of diffusion by comparing the findings of the 1991 and 1993 surveys while the last part evaluates the survey findings in terms of the theoretical models of the diffusion process referred to in the introductory chapter of this book.

It should be noted that these findings are based on the responses obtained by means of a comprehensive telephone survey of all 514 authorities in Great Britain. A telephone based survey strategy was adopted because of the large number of postal questionnaires which were being circulated to local authorities and a resulting concern about the size of the response. The 100% response rate achieved supports the adoption of this method and also enabled respondents to give valuable subsidiary information during the interviews. The method also removes the ambiguity which exists in some surveys about respondents' perceptions of the definition of GIS, as the capabilities of the software and the precise nature of its use in the host authority are often related to operational definitions. The research has adopted a broad interpretation of GIS which includes automated mapping and facilities management type systems but excludes thematic mapping and computer aided design packages. One respondent, generally the project manager, was interviewed with respect to each separate system present within a particular authority.

THE STRUCTURE OF BRITISH LOCAL GOVERNMENT

The structure of local government in the United Kingdom takes three distinct forms. Scotland and Northern Ireland each have their own types of operation, while England and Wales share a common structure. As a result of the special circumstances in Northern Ireland the functions of local government in this case have largely been taken under the control of central government. It is not therefore appropriate to include Northern Ireland within the study.

The system with which the majority of the population in Great Britain was familiar at the time of the survey was that found in England and Wales. In this case a dual system was in operation whereby metropolitan and non-metropolitan areas each had separate structures. For the metropolitan areas, which comprise the six large conurbations and London, there were 35 metropolitan districts and 33 London boroughs. These are all subsequently referred to as metropolitan districts. Since the abolition of the upper tier metropolitan counties in 1986 these multi-purpose unitary authorities have worked alongside a range of ad hoc bodies which were introduced at the time to coordinate metropolitan wide activities such as passenger transport and the police and fire services. In contrast, in the non-metropolitan areas at the time of the survey, there was a two tier structure consisting of 333 districts and 47 counties. These authorities will subsequently be referred to as shire districts and shire counties. At the time of the survey the Scottish system made no distinction between metropolitan and non-metropolitan areas having a two-tier structure of 53 districts and 9 regions. There were, however, three unitary authorities for the Island communities which depart from the general structure.

The division of functions in the non-metropolitan areas results in districts largely taking over the responsibility for local scale services such as housing, public health, refuse collection and local land use planning. In England and Wales the higher level authorities had responsibility for the provision of services such as education, roads and transport, strategic land use planning, the emergency services and social services. In Scotland the regions had rather a broader role, including such activities such as supplying water. As a result of this division of responsibilities and the difference in size, the counties and regions command much larger budgets than the shire and Scottish districts.

In contrast to many other countries in Europe, British local authorities have responsibility for a relatively limited range of activities (Norton, 1991). For example, provision of most health care facilities, and in England and Wales, water supply, are undertaken by either unelected regional bodies or privatized companies.

Another distinctive feature of the British local government system is the size of the basic unit of local authorities. The average populations of the Scottish, shire and metropolitan districts at the time of the 1991 census were 94 300, 96 700, 253 600, and even the smallest shire district of Teesdale had a population of 24 068. In contrast, the populations of many English counties and Scottish regions are smaller than the highest level sub-national government in many European countries, with average populations of 685 000 and 547 400 respectively in 1991. This reflects the relative absence of regional government in Great Britain.

The appropriateness of the two-tier structure of local government in the shire counties and Scotland is currently a cause of concern for the Government who initiated a full review of arrangements in 1991. The process in England is being undertaken on a county by county basis and the precise outcomes are not yet fully clear. However, it appears likely that the result of this review will be a patchwork of unitary authorities for the large urban areas and a mixture of single and two tier government for the rest. In the case of Scotland proposals are being implemented which will remove the upper tier of local government. In the case of Wales all but one of the counties are being abolished and replaced by unitary authorities.

The effects of this review on the adoption and take up of GIS in British local government have been mixed. Some respondents in the 1993 survey indicated that their authorities were deferring decisions to acquire GIS until the uncertainties about their futures were resolved. However, there were others who claimed that their authorities had brought forward the acquisition of GIS facilities with a view to strengthening their bargaining positions in future negotiations.

DIGITAL TOPOGRAPHIC DATA AVAILABILITY

Britain's national mapping agency, the Ordnance Survey, provides a high quality, comprehensive, up to date mapping service for a very wide range of users in England, Wales and Scotland but not Northern Ireland which has its own mapping agency (Rhind, 1991). In recent years there has been a marked increase in the availability of digital topographic data, particularly at the large scales that are used for many local government tasks. By the end of 1995 it is envisaged that all 57 564 1:1250 scale tiles of urban areas and all 156 700 1:12500 scale tiles of rural areas data will be available in digital format to users. The Ordnance Survey is also developing a variety of additional geographic information services in collaboration with other government agencies in the private sector. A good example of these services is the development of Addresspoint which locates all 25 million postal addresses in Great Britain to 1 m resolution. This product makes use of the Postal Address File developed by the Royal Mail and the street centre line files produced by the Ordnance Survey itself.

The Ordnance Survey is both a government department and a Executive Agency. This means that it operates within the financial regime that has been applied to more than 60 other government agencies by the British Government whereby targets are set annually and performance is measured against them (Rhind, 1993). Of particular importance in this respect are the financial targets that have been set by Government. In 1991/92 the Ordnance Survey recovered 68% of all its costs, including accommodation, by the sale of goods and services and the present Government has made it clear that it wishes the Ordnance Survey to reach full cost recovery by 1997. Under these circumstances, large scale digital data have become increasingly available to local government users at a time of considerable uncertainty as to their likely cost.

For this reason, the Service Level Agreement which was reached between the Ordnance Survey and the local authority associations in March 1993 regarding the acquisition of topographic data in digital or geographic format is of considerable importance. This marks an important departure from previous agreements whereby individual local authorities or departments negotiated their own terms with the Ordnance Survey for the purchase of such data. The potential impact of this agreement on the diffusion of automated mapping and GIS facilities in Great Britain is considerable. It should be noted that the survey whose findings are reported in the next two sections took place between June and September 1993, that is between 3 and 6 months after the agreement was reached and before the full potential was realized. Some indication of the full potential of this agreement can be seen from the returns after the first year of operation in April 1994 which show that 460 local agreements had been signed by local authorities at that stage. Some 289 of these agreements, or 62.8% of the total, were concerned with large scale digital map data. This is nearly double the number of authorities which had GIS facilities in 1993 (Campbell *et al.*, 1994, p. 37).

Unlike most European countries property rights are transferred by deeds of conveyance under the legal system of Great Britain (Lievesley and Masser, 1993). Recently, however, considerable progress has been made in promoting the registration of land titles in England and Wales. By 1994 there were over 15 million titles registered in HM Land Registry and nearly ten million of them had been converted to computer format (Manthorpe, 1994). As might be expected the growing availability of land registration data in digital format has prompted a number of projects exploring the feasibility of developing an integrated National Land Information System [see, for example, Dale (1994)]. Of particular interest in this respect is the pilot project which is taking place in Bristol at the present time as a collaborative venture between the Valuation Office, the Land Registry, the Ordnance Survey and the City Council.

GIS ADOPTION

The findings of the survey summarized in Table 4.1 give an overall picture of the state of local authority plans for GIS in summer 1993. From this it can be seen that 149 out of the 514 authorities or 29% of the total in Great Britain had GIS facilities and that a further 50 authorities had firm plans to acquire them within a year. Many of these have already set up working parties and undertaken feasibility studies to evaluate GIS. As a result, it is likely that about two authorities in five in Great Britain will have GIS facilities at the end of 1994.

Some 139 authorities were considering the acquisition of GIS at the time of the survey and only 176 authorities had no plans to introduce GIS. The main reasons given by the latter were the uncertainties associated with the Government's proposals to reorganize local government and lack of finance. However, a number of authorities reported that they had set up working parties to consider GIS and decided not to proceed as a result of their negative recommendations.

The findings demonstrate the high level of awareness of GIS in local government circles in Great Britain even amongst authorities that had no plans to invest in such systems. However, there were marked differences between different types of authority and also between different regions with respect to GIS adoption. Table 4.2 shows

Table 4.1 Plans for GIS in local authorities in Great Britain

	1993	
Plans for GIS	Number	%
Already have GIS facilities	149	29.0
Plans to acquire GIS within 1 year	50	9.7
Considering the acquisition of GIS facilities	139	27.0
No plans to introduce GIS	176	34.2
Total	514	99.9

Table 4.2 Plans for GIS by type of local authority

	With GIS	
Type of authority	Number	%
Shire districts	61	18.3
Metropolitan districts	34	49.3
Shire counties	43	91.5
Scottish districts	5	9.4
Scottish regions	6	66.7
Scottish islands	0	0.0
Total	149	29.0

the number of authorities who have already acquired a GIS by local authority type. From this it can be seen that the highest level of GIS adoption is at the shire county and Scottish region levels where 43 out of the 47 counties or 91% of the total and six out of the nine regions already had GIS facilities.

In contrast only 61 out of the 333 shire districts (18.3%), 5 out of the 53 Scottish districts (9.4%) and none of the three Scottish Island authorities had acquired GIS by summer 1993. As might be expected the level of adoption in the metropolitan districts fell mid-way between those of the shire counties and Scottish regions on the one hand and those of the shire districts and Scottish districts on the other. Table 4.2 shows that 34 out of the 69 metropolitan districts (49.3%) had GIS at the time of the survey. It is important to note, however, that despite the relatively low level of adoption in shire districts this type of authority had the largest absolute number of authorities with GIS.

Table 4.2 also shows that levels of adoption in the Scottish regions and Scottish districts are generally lower than those in England and Wales. The extent of these regional differences is further explored in Table 4.3 with reference to the North/South divide. This indicates that, in overall terms, 32% of authorities in the South which includes London, the South East region together with East Anglia, the East and West Midlands and the South West had GIS facilities as against only 24% of authorities in the North which includes Wales, Yorkshire and Humberside, the Northern and

Table 4.3   Percentage of local authorities with GIS facilities by type and region

Type of Authority	% South with GIS	% North with GIS	% All authorities
Shire districts/			
Scottish districts	21.0	10.9	17.0
Metropolitan districts	52.5	44.8	49.3
Shire counties/			
Scottish regions	90.3	84.0	87.5
All authorities	32.2	24.3	29.0

North Western regions as well as Scotland. As might be expected these variations were least pronounced with respect to the shire counties and Scottish regions where GIS adoption is approaching 100% cent and most pronounced with respect to the shire districts and Scottish districts where GIS adoption is still under 20%. In the former case the ratio south to north in percentage terms is only 90:84 whereas it is 21:11 in the latter. Generally differences between the metropolitan districts are closer in this case to those of the shire counties and the Scottish regions with a ratio of 53:45.

The picture of GIS adoption in summer 1993 that emerges from these survey findings is very clear. Overall levels of adoption were highest in the shire counties and the Scottish regions where nearly all authorities had GIS facilities and lowest in the Scottish shire districts and Scottish districts where less than 20% of authorities had GIS. About half the metropolitan districts had GIS at the time of the survey. Levels of adoption were also high in the southern and eastern regions of Britain and lower in the northern and western regions. The extent of these regional variations was particularly marked with respect to the shire districts and Scottish districts. In this case the probability that a shire district in the south had GIS was nearly twice as high as that for a shire district or Scottish district in the northern part of Great Britain.

AUTHORITIES ADOPTING GIS

The previous section dealt with the number of authorities with GIS facilities. To analyse GIS usage it is necessary to consider the number of systems purchased by these authorities and their configuration. A system is regarded as a distinct piece or combination of software which one or more departments within a local authority are implementing. For instance, a situation where several departments are developing separate applications based on the same software is considered as one system.

Table 4.4 shows that a total of 195 separate systems had been purchased in 149 authorities which had acquired GIS facilities by summer 1993. Shire counties and Scottish regions were most likely to have more than one system within an authority with 75 systems in 43 authorities and 11 systems in 6 authorities respectively. In contrast only one of the 61 shire districts and two of the seven Scottish districts had more than one system.

Table 4.5 shows that more than 70% of all authorities had purchased systems since the beginning of 1990. The peak year for local authorities to move into GIS

Table 4.4 Level of GIS adoption by authorities and systems

Type of authority	No. of authorities Possessing GIS	Number of GIS
Shire districts	61	62
Metropolitan districts	34	40
Shire counties	43	75
Scottish districts	5	7
Scottish regions	6	11
Total	149	195

Table 4.5 Length of experience with GIS technologies

Year	Authorities		Systems	
	Number	%	Number	%
Before 1986	5	3.4	7	3.6
1986	1	0.7	2	1.0
1987	3	2.0	5	2.6
1988	15	10.1	16	8.2
1989	12	8.1	15	7.7
1990	35	23.5	39	20.0
1991	26	17.4	36	18.5
1992	30	20.1	46	23.6
1993[a]	22	14.8	29	14.9
Total	149	100.1	195	100.1

[a] Up to September only.

for the first time was 1990 itself when 35 authorities acquired facilities. Since that time the number of authorities adopting GIS each year has declined slightly. Table 4.5 also shows that over three-quarters of all systems had been purchased since the beginning of 1990 and that the peak year for system acquisition was 1992. It should be noted, however, that the figures for 1993 are not complete because of the timing of the survey.

The differences noted above regarding the number of systems in various types of authority reflect both the resources at the disposal of the authority and the organizational arrangements within that authority for carrying out their statutory duties. At the same time they give some indication of the diversity of approaches that are being adopted to GIS implementation in British local government. From Table 4.6 it can be seen that 105 out of the 195 systems were at the disposal of a single department while the other 90 systems were shared by more than one department. Departmental systems were particularly common in shire counties and Scottish regions where they accounted for 59 out of the 86 systems in use. In contrast systems involving more than one department accounted for three out of every five facilities in the shire districts and Scottish districts. In the metropolitan districts the

Table 4.6 Approach to GIS implementation

Type of authority	Single department		More than one department		Total
	Number	%	Number	%	
Shire and Scottish districts	27	39.1	42	60.9	69
Metropolitan districts	19	47.5	21	52.5	40
Counties and regions	59	68.6	27	31.4	86
Total	105	53.8	90	46.2	195

Table 4.7 Number of departments involved in multi-departmental GIS facilities

Number of departments	Shire/Scottish districts %	Metropolitan districts %	Counties and regions %	Total %
2	19.0	23.8	40.7	26.7
3	35.7	19.0	29.6	30.0
4	9.5	14.3	14.8	12.2
5	19.0	14.3	0.0	12.2
More than 5	9.5	28.6	14.8	15.6
All departments	7.1	0.0	0.0	3.3
	($n = 42$)	($n = 21$)	($n = 27$)	($n = 90$)

distribution of single and multi-departmental systems was relatively evenly balanced with 19 single as against 21 multi-departmental systems.

In no way can multi-departmental systems be equated automatically with corporate systems of the kind described in some of the literature [see, for example, Bromley and Coulson (1989), James and Pope (1993), Mahoney (1989), Moore (1994)]. Table 4.7 shows that the majority of multi-departmental facilities involved only two or three departments, and less than one GIS facility in five involved five or more departments. Facilities involving two or three departments were particularly popular in the shire counties and the Scottish regions where they accounted for over 70% of all multi-departmental facilities. However, even in the shire districts and Scottish districts facilities involving two or three departments accounted for over half the total in this category. Conversely, only 7% of facilities in the shire and Scottish districts and no facilities at all in the metropolitan districts or the shire counties and Scottish regions involved all departments in the authority.

Table 4.8 shows that 404 separate departments were involved in the 195 GIS facilities in summer 1993. A quarter of these departments were planning or development departments. Another quarter was accounted for by highways and estates departments. IT and technical services were also well represented in this as were legal services, parks and recreation and the Chief Executive's Department. The 'others' category which accounted for 61 out of the 404 departments included big

Table 4.8 Departments involved in GIS facilities

Departments	Shire/Scottish districts	Metropolitan districts	Counties and regions	All authorities
Planning/development	49	25	32	106
Highways/engineers/ surveyors	16	15	28	59
Estates	19	13	19	51
IT/computer services	14	5	14	33
Combined technical services	12	7	12	31
Legal and related services	19	4	0	23
Parks/recreation	13	5	2	20
Chief executive	6	9	5	20
Others	24	27	10	61
Total	172	110	122	404

Table 4.9 Lead departments in multi-departmental facilities

Departments	Shire/Scottish districts	Metropolitan districts	Counties and regions	All authorities
Planning/development	17	11	4	32
IT/computer services	4	4	5	13
Combined technical services	4	1	5	10
Highways/engineers	0	0	5	5
Chief executive	3	0	1	4
Legal and related services	3	0	0	3
Others	4	2	2	8
No lead department	7	3	5	15
Total	42	21	27	90

spending departments such as education (10 cases), housing (9), and social services (5), as well as environmental health (14) and buildings and works (4).

In more than a third of multi-departmental facilities, as Table 4.9 shows, planning was the lead department. Only IT/computer services and combined technical services were lead departments in 10 or more facilities and in 15 cases there was no single lead department. Planning was particularly important as the lead department in shire district and Scottish district applications and at the metropolitan district level. At the county level, planning together with highways and the two technical services departments were all lead departments in four or five instances.

Table 4.10 shows that planning departments also accounted for a third of all single department GIS facilities and over half at the shire district/Scottish district level.

Table 4.10   Single department GIS

Departments	Shire/Scottish districts	Metropolitan districts	Counties and regions	All authorities
Planning/development	15	8	12	35
Highways/engineers	2	2	12	16
Emergency services	0	0	15	15
Combined technical services	4	2	7	13
Estates	2	2	7	11
Others	4	5	6	15
Total	27	19	59	105

Other major users of single department facilities were highways/engineering and emergency services especially in the shire counties and the Scottish regions and combined technical services and estates.

The overall picture of GIS system development that emerges from the survey findings is one of considerable diversity reflecting the specific demands of particular types of authority and different approaches to organization and management. The findings generally suggest that GIS in British local government is decentralized and largely a bottom-up activity and that centralized systems are generally in the minority. Single departmental systems predominate in shire county and Scottish region GIS facilities while multi-departmental systems were the most common feature of shire district and Scottish district facilities. The tendency towards fragmentation of GIS in the counties which was evident in the number of systems in use is further reinforced by an emphasis on two or three departmental systems even in cases where multi-departmental systems are being implemented.

In all types of authority, planning and development departments were the most frequently cited department. Planning was also the lead department in over a third of the multi-departmental facilities. The predominance of planning reflects the importance these departments attached to geographic information as well as their traditional responsibilities for meeting the cartographic needs of local authorities. The pre-eminent position of planning was challenged only by highways/engineering departments in the shire counties and Scottish regions with respect to single department facilities. Given this emphasis on operational departments it is not surprising to find that IT/computing services together with other central departments such as legal and related services and the Chief Executive's departments were lead departments only in a relatively small number of multi-departmental applications.

Choice of technology

Table 4.11 provides a breakdown of the software packages acquired by local authorities. It should be noted that systems with GIS capabilities that are largely being used to perform activities such as CAD have been omitted from this analysis. It should also be noted that where packages had been purchased to provide specialist

Table 4.11 Software adopted for GIS work

Software	Shire/Scottish districts %	Metropolitan districts %	Counties and regions %	All authorities %
Arc/Info	10.1	27.3	29.1	22.0
Wings	5.8	—	11.6	7.2
MapInfo	8.7	7.5	4.6	6.7
Axis	13.1	10.0	—	6.7
Alper GIS	10.1	10.0	1.2	6.2
G-GP	5.8	15.0	1.2	5.6
X Assist	—	—	11.6	5.1
GDS	8.7	5.0	1.2	4.6
Others	37.7	25.0	39.5	35.9
Total	100.0	100.0	100.0	100.0
	($n = 69$)	($n = 40$)	($n = 86$)	($n = 195$)

GIS facilities to supplement an existing system it is the main system which is recorded in the table.

A striking feature of the survey findings is the pre-eminent position of Arc/Info (ESRI) as the leading GIS software package in British local government with a market share of about 22%. Arc/Info was used mainly by the shire counties and Scottish regions together with the metropolitan districts. The market leader for the shire districts and Scottish districts was Axis (a system developed by a British company in Northampton) with 13.1% of the total number of systems. Other software packages used extensively at the shire district and Scottish district level were Alper GIS (another system developed by a British company based in Cambridge which is now part of Sysdeco) and Arc/Info. Arc/Info, Axis and Alper GIS were also used extensively by the metropolitan districts as was G-GP (another British system). The latter is largely due to the development of the software for the Research and Intelligence Section of the old Greater London Council and its subsequent popularity with the London boroughs. Two packages which were widely used at the shire county and Scottish regional level but not elsewhere were Wings (another British system) and X-Assist (a specialist system developed for environmental monitoring).

Table 4.11 also highlights the variety of software that was in use at the time of the survey. The 'others' category accounted for over a third of all applications. This included more than 30 additional software packages as well as some in house home grown products. The greatest range of systems was found at both the shire county and the Scottish regional level and in the shire districts and Scottish districts.

In the survey respondents were asked questions about the hardware used to support GIS. In cases where systems had several means of access, for instance mainframe terminals and microcomputers, it was the highest level of capability, that is the mainframe component which was used in the analysis.

Overall, as Table 4.12 shows, mainframe installations accounted for only one facility in 12 in Summer 1993. The dominant hardware platform for GIS in all types of authority was the workstation followed by the PC. These two hardware types accounted for 55.4 and 32.3% of all applications respectively.

Table 4.12   Hardware adopted for GIS work

Hardware	Shire/Scottish districts %	Metropolitan districts %	Counties and regions %	All authorities %
Workstation	55.1	60.0	53.5	55.4
Micro	34.8	22.5	34.9	32.3
Mainframe	10.1	7.5	5.8	7.7
Mini	—	10.0	1.2	2.6
Other	—	—	4.6	2.0
Total	100.0	100.0	100.0	100.0

Benefits and Problems Associated With GIS

Respondents were asked to rank in order of importance three sets of benefits and problems associated with GIS. They were then asked to describe in more detail the group of benefits and problems they had ranked first with reference to a more detailed list of topics. Provision was also made for issues not identified in the questionnaire to be raised. Although the findings from this part of this survey are particularly dependent upon the perceptions of the individuals interviewed they nevertheless enable some general issues to be identified.

Table 4.13 shows that the most important benefit associated with GIS was better information processing. More than 60% of all respondents ranked this first. Their main reasons for giving this first place were improved data integration, increased speed of data provision, better access to information and an increased range of analytical and display facilities.

The next most important factor in the ranking was better quality decisions which was placed first by 20.8% of respondents. Better quality decisions were felt to be particularly important by shire counties and Scottish regions. The most important reasons given for this ranking were related to operational and managerial decision making rather than policy making matters. The third factor, general savings, was placed first by only 11.4% of respondents. In this case the most important savings were associated with reductions in time rather than in cash or staff.

Table 4.14 shows that respondents were more or less evenly divided between organizational, technical and data related issues when it came to ranking problems.

Table 4.13   Most important benefits associated with GIS

Benefits	Percentage of all authorities in 1993
Improved information processing facilities	61.4
Better quality decisions	20.8
General savings	11.4
Others	6.4
	($n = 98$)

Table 4.14 Most important problems associated with GIS

Problems	Percentage of all authorities in 1993
Technical	29.0
Organizational	27.2
Data related	26.3
Others	7.6
No problems	9.8
	$(n = 195)$

A wide variety of technical issues were raised by respondents who ranked technical matters first in their list of problems. The most commonly cited technical problem was hardware reliability followed by lack of systems compatibility. Other problems cited by respondents included software capability, lack of user friendliness and difficulties with vendors.

The most important organizational problems perceived by respondents were the poor quality of managerial structures for implementing GIS followed by shortage of skilled staff and the lack of encouragement from senior staff. In several cases staff resistance also presented a significant problem.

By far the most commonly cited data related problem was the cost of data capture. Respondents were not only concerned about the cost associated with data capture but also the amount of time required for this task. Other problems cited by some respondents included lack of compatibility between data sets, poor quality and difficulties with Ordnance Survey data. Problems of lack of compatibility were predominantly cited by respondents from shire counties and Scottish regions.

THE DYNAMICS OF DIFFUSION 1991–93

The 1993 survey was carried out in the same format as that used in the earlier survey undertaken $2\frac{1}{2}$ years earlier. Because of this it is possible to directly compare the findings of the two surveys to highlight the dynamics of GIS diffusion within British local government. Some of the main similarities and differences between the surveys are summarized in Tables 4.15 and 4.16 with respect to the headings used above.

From Table 4.15 it can be seen that the number of authorities with GIS had increased by 75% over the $2\frac{1}{2}$ years from 85 to 149. As might be expected the rate of increase was lowest in the shire counties and Scottish regions where adoption levels were approaching 100% and highest in the shire districts and Scottish districts where adoption levels were still below 20%. In the latter the number of authorities with GIS had more than doubled since the time of the 1991 survey.

These figures highlight the volatile nature of GIS diffusion in British local government over the last few years. If the expectations of the 50 additional authorities who indicated that they had firm plans to acquire GIS within 1 year have since been realized, the overall level of GIS adoption would have risen by a further 35% over 1993 levels by the end of 1994. In practice, even this could be an underestimate given the boost to GIS adoption given by the Service Level Agreement reached between

Table 4.15   Summary of main changes between 1991 and 1993 with respect to GIS adoption and system development

	1991	1993
Adoption		
No. of authorities with GIS	85	149
Shire counties/Scottish regions	35	49
Metropolitan districts	22	34
Shire districts/Scottish districts	28	66
South north ratio (% adoption)	20:11	32:24
System Development		
No. of systems in local authorities	98	195
Average length of experience (years)	2.51	2.94
Single multiple departmental ratio	49:51	54:46
Average no. of departments in multi-departmental applications	4.18	3.81
% planning as lead department in multi-department applications	27.0	35.6
% planning of single departmental applications	22.9	33.3

Table 4.16   Summary of main changes between 1991 and 1993 with respect to GIS technology and perceived benefits and problems

	1991	1993
Technology		
Software packages (% total)		
Arc/Info	22.4	22.0
Wings	4.1	7.2
Axis	7.1	6.7
MapInfo	3.6	6.7
Alper Records	12.2	6.2
G-GP	7.1	5.6
GFIS	10.2	3.0
Hardware configurations (% total)		
Workstation	40.8	55.4
Mainframe	25.5	7.7
Benefits (% total)		
Improved information processing	60.5	61.4
Better quality decisions	31.5	20.8
Problems (% total)		
Technical	28.4	29.0
Organizational	28.4	27.2
Data related	34.0	26.3

Britain's national mapping agency, the Ordnance Survey, and the local authority associations in March 1993 referred to earlier in the chapter.

Table 4.15 also shows that, as overall levels of adoption rise, regional variations tend to be reduced. Whereas the probability of an authority in the southern part of Britain adopting GIS was nearly two to one in 1991, by 1993 it was only four to three. However, these overall figures conceal important differences between authority types. Between 1991 and 1993 the south:north ratio actually increased slightly with respect to the smaller shire districts and Scottish districts whereas it was already approaching one to one for the shire counties and Scottish regions.

During the $2\frac{1}{2}$ years between the two surveys the number of systems in local authorities had effectively doubled from 98 to 195. However, because of the rapid increase in acquisitions, the average length of experience with these systems had risen only from 2.51 to 2.94 years during the period. Consequently, for the vast majority of British local authorities, GIS is still a new technology and they are still in the process of building-up operational experience.

The growing fragmentation of GIS in British local government evident in the number of authorities which have acquired more than one system especially at the shire county and Scottish region level is also apparent in the configuration of systems. Table 4.15 shows that the ratio of multi-departmental to single departmental facilities was reversed between 1991 and 1993 from a slight majority in favour of multi-department facilities to a clear 54:46 majority in favour of single department facilities in 1993. It is also worth noting that the average number of departments involved in multi-departmental facilities went down from 4.18 to 3.81 over this period.

Given the enormous diversity of local government operations it is surprising to find that planning and development departments have actually consolidated their position as GIS leaders within local government since 1991. In 1993 planning and development departments acted as the lead department for 35.6% of multi-departmental facilities as against only 27% in 1991 and planning and development departments were also responsible for a third of all single department applications in 1993 as against only 22.9% in 1991.

Table 4.16 summarizes the main changes that took place in the $2\frac{1}{2}$ years between the 1991 and 1993 surveys with respect to technology and perceived benefits and problems. From this it can be seen that there was dramatic fall in the proportion of mainframe installations during this period and an increased dominance of workstation based facilities. In contrast, Arc/Info has retained its market share over this period despite the growing number of software packages on the market. In practice Arc/Info has consolidated its position at a time when the market shares of the other leading packages have declined. This is particularly evident in the case of GFIS whose market share slumped from 10.2% in 1991 to only 3% in 1993.

It is also worth noting that several packages increased their market shares between 1991 and 1993. The main gainers were Wings and MapInfo whose shares rose from 4.1% to 7.2% and 3.6% to 6.7% respectively. It should be borne in mind, that in addition to first time purchases, a number of authorities reported that they had switched software during this period. The most commonly cited switch was from GFIS to Arc/Info.

By comparison with these changes there were no dramatic shifts in the benefits and problems associated by respondents with GIS. Improved information processing was perceived as the main benefit by three out of five respondents in both 1991 and 1993. However the proportion of respondents who chose better quality decisions as the main benefit declined from over 30% to 20% during this period. This may reflect the increased number of district level as against county level facilities.

The only discernible shift in perceived problems was the fall in the number of respondents who saw data related issues as the main problems associated with GIS. This probably reflects the extent to which the lead time for data capture has been reduced as authorities build up their operational experience as well as the impact of the Service Level Agreement negotiated by the local authority associations with the Ordnance Survey.

THE DIFFUSION PROCESS

From Figure 4.1 it can be seen that the rate of adoption of GIS by British local authorities up to 1991 is broadly similar to the first section of the S-shaped curve described in Chapter 1 while the subsequent increase in the number of authorities adopting GIS has all the characteristics of the second section of the curve. However, the point at which the curve levels off again to the third section has not yet been reached and therefore it is not possible to predict the final outcome of the diffusion process. It is worth noting in this respect the reference in the section on adoption to the fact that a number of authorities reported that they had set up working parties to consider GIS and decided not to proceed as a result of the negative recommendations. Findings such as these raise important questions as to the inevitability of GIS adoption in all British local authorities.

Although the survey evidence is essentially cross sectional in nature it nevertheless contains many findings which support the assumptions underlying both the hierarchical and core-periphery models described in Chapter 1. As far as the hierarchical model is concerned over 90% of English counties which form the top tier of local government already had GIS facilities at the time of the 1993 survey. In contrast only half the metropolitan districts that form the next tier of local government in terms of size had acquired GIS and less than one in five shire districts and only one in ten Scottish districts that make up the lowest tier of local government had such a facility.

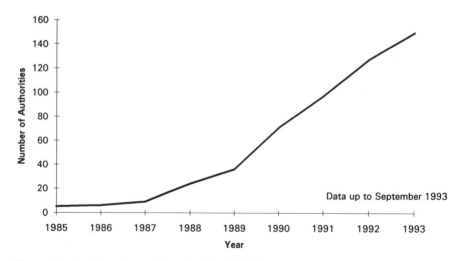

Figure 4.1   Number of authorities adopting GIS, 1993.

A more surprising finding from the surveys is the extent to which the evidence supports the core-periphery model of diffusion even in a relatively small country such as Britain where there are well established institutions in existence to disseminate new ideas to all parts of the country at the same time. Nevertheless, as Table 4.3 shows, the probability of local authorities of all types having GIS facilities was higher in the south of Britain than in the north and Scotland. As might be expected, given the findings of the hierarchical model, these differences were least marked at the shire county and Scottish region level and most marked at the shire district and Scottish district levels. Consequently, as Masser's (1993) multi-nomial logit analysis of the findings of the 1991 survey showed, more than half the variance in the data set as a whole can be explained by three variables: type of authority, population size and location in the core or periphery.

CONCLUSIONS

The findings of the 1993 survey reported in this paper provide a comprehensive overview of GIS diffusion in British local government. They indicate both the extent to which GIS has already penetrated the local government sector and the diversity of arrangements that are being made for GIS implementation. The present analysis has concentrated largely on GIS adoption. A detailed evaluation of the implementation of experience of 12 case study authorities can be found in another work by the authors (Campbell and Masser, 1995).

The volatile nature of the GIS diffusion process is highlighted when the findings of the 1993 survey are compared with those from an earlier survey carried out by the authors using the same methodology. During the $2\frac{1}{2}$ years between the two surveys the number of systems had effective doubled. This period also saw the virtual demise of mainframe GIS. Because of the number of new acquisitions, however, GIS is still a very new technology in most authorities and the average length of experience is generally less than 3 years.

Given the pace of recent change and the likelihood of continuing rapid developments in the future, the findings of this analysis demonstrate the importance of regular and systematic monitoring of the diffusion of GIS in British local government. Further surveys of this kind are essential not only to keep track of events but also to identify new issues that are emerging within the local government GIS community.

The analysis also draws attention to some of the distinctive features of the institutional context in British local government. Of particular importance in this respect are the size of the lower tier authorities by comparison with their European counterparts and the comprehensive range of large and small scale topographic mapping services provided by the Ordnance Survey. Features such as these must be borne in mind when comparing the findings for British local government with those for other European countries in subsequent chapters of this book.

ACKNOWLEDGEMENTS

The authors are grateful to the Economic and Social Research Council (ESRC) and the Natural Environment Research Council (NERC) for funding the initial development of this project as part of their joint programme on Geographic

Information Handling. They also wish to thank the Local Government Management Board who funded the 1993 survey whose findings are reported in this chapter. Some sections of this chapter have already been published in another book by the authors on GIS and Organizations. The authors are grateful to the publishers of this book, Taylor and Francis, for granting permission to include these extracts in this chapter.

REFERENCES

BROMLEY, R. and COULSON, M., 1989. The value of corporate GIS to local authorities: evidence of a case study in Swansea City Council, *Mapping Awareness*, **3**(5), 32–5.

CAMPBELL, H. and MASSER, I., 1992. GIS in local government: some findings from Great Britain, *International Journal of Geographic Information Systems*, **6**, 529–46.

CAMPBELL, H. and MASSER, I., 1995. *GIS and Organisations*, London: Taylor and Francis.

CAMPBELL, H., MASSER, I., POXON, J. and SHARP, E., 1994. *Monitoring the Take Up of GIS in British Local Government*, Luton: Local Government Management Board.

DALE, P., 1994. Towards a national land information system, in Green, D., Rix, D. and Cadoux Hudson, J. (Eds) *Geographic Information 1994*, pp. 233–6, London: Taylor and Francis.

JAMES, P. and POPE, R., 1993. Corporate GIS in local government: the Horsham experience, *Mapping Awareness*, **7**(4), 28–30.

LIEVESLEY, D. and MASSER, I., 1993. An overview of geographic information in Europe Part I, *Mapping Awareness*, **7**(10), 9–12.

MAHONEY, R., 1989. Should local authorities use a corporate or departmental GIS? *Mapping Awareness*, **3**(2), 57–9.

MANTHORPE, J., 1994. England and Wales, *Geodetical Info Magazine*, **8**(2), 42–5.

MASSER, I., 1993. The diffusion of GIS in British local government, in Masser, I. and Onsrud, H.J. (Eds) *Diffusion and Use of Geographic Information Technologies*, pp. 99–115, Dordrecht: Kluwer.

MASSER, I. and CAMPBELL, H., 1993. The impact of GIS on local government in Britain in Mather, P.M., (Ed.) *Geographic Information Handling: Research and Applications*, pp. 273–86, Chichester: John Wiley.

MOORE, K., 1994. GIS implementation in local government: the central approach, in *Proceedings of the Association for Geographic Information Conference*, pp. 20.1.1–20.1.5, Rickmansworth: Westrade Fairs Ltd.

NORTON, A., 1991. Western European local government in comparative perspective, in Batley, R. and Stoker, G. (Eds) *Local Government in Europe: Trends and Developments*, pp. 21–40, London: Macmillan.

RHIND, D., 1991. The role of the Ordnance Survey in Britain, *Cartographic Journal*, **28**, 188–99.

RHIND, D., 1993. Policy on the supply and availability of Ordnance Survey information over the next five years, in *Mapping Awareness*, **7**(1) 37–41.

CHAPTER 5

Germany: a federal approach to land information management

HARTWIG JUNIUS, MICHAEL TABELING and MICHAEL WEGENER

INTRODUCTION

The diffusion of geographical information systems (GIS) in local government in Germany has been handicapped by (1) the federal organization of government in Germany giving a relatively large degree of autonomy to local governments and hence little incentive for cooperation, (2) the fact that most commercially distributed GIS products are of US origin and require English-language capabilities, and (3) a general drift of local planning away from rationalist, comprehensive and long-range approaches towards incremental, discursive ones based on informal rather than on formal information (see Wegener and Junius, 1991, 1993).

This paper summarizes the present state of development of geographical information systems in Germany, reports on recent efforts to improve data integration within the federal system and points to current problems. It finally presents a survey of GIS diffusion in German cities following the methodology developed at the University of Sheffield. In the survey 86 German cities with a population of more than 100 000 were asked about their GIS experience and plans.

The survey confirmed the delay in GIS diffusion in local governments in Germany. Of the 80 cities responding to the survey 70 have some experience with GIS technology, but only 45 have to date acquired full-scale integrated GIS. On the level of GIS applications the gap between actual and potential GIS use is even larger: less than one fifth of all potential GIS applications have been implemented or are in the planning stage. Surprisingly, not the largest cities but the cities in the upper mid-range of city sizes are the most active in adopting GIS technology. However, where GIS applications are being implemented, spatial data bases and spatial reference systems tend to be well integrated between different map scales and departments and agencies. As all cities presently without a GIS stated their intention to adopt GIS technology, it can be expected that GIS diffusion in German cities will make rapid progress in the near future.

ALB/ALK: A DIGITAL CADASTRE

Basic land information in Germany is maintained in a dual mode (see Wegener and Junius, 1991). There are two registers: the cadastre (Liegenschaftskataster) and the land register (Grundbuch). Because of the federal structure of Germany, surveying tasks are performed by the Länder and governed by surveying and cadastral laws. In the former GDR, district authorities were responsible for the cadastre but neglected it for political reasons. After the unification, the new Länder followed the west German example.

Because in the 19th century property taxes were the principal income of local governments, some of the larger territories developed land registers in which individual lots were listed by area, soil quality and owner. Later the land registers also contained information on titles such as mortgages. The systematic inventory of all properties in the form of cadastral registers and maps based on surveying property boundaries was introduced in the 1820s. The installation and maintenance of the cadastre was assigned to surveying departments as a permanent task.

With the decreasing role of property taxes for local government income and the growing importance of land as an economic asset, the principal function of land information became the registration of property rights. For this the cadastre was complemented by a land register containing information on land ownership titles.

This led to the dual mode of land information existing in Germany today. The cadastre contains information on the physical characteristics of properties such as boundary, label, size, land use, location, etc. The land register, on the other hand, contains information on the ownership of properties and property rights. The two registers are linked by a common referencing system based on a hierarchical numeric code for each property, and each change in one of them affects the other. This makes updating cumbersome and slow but guarantees a high level of reliability, as the cadastre participates in the official status of the land register. Liegenschaftskataster and Grundbuch might, directly or indirectly, be linked to the files of the local population register, planning and budgeting departments and the Commission for Land Value Assessment.

In 1971 a working party of representatives of the Länder submitted a proposal for automating both registers. The first project started was the automated cadastral register (Automatisiertes Liegenschaftsbuch), in short ALB, designed as a standard database management system (DBMS). The second project, started in 1976, is the automated cadastral map (Automatisierte Liegenschaftskarte), or ALK. That the two projects were conducted separately was due to the fact that data structures to efficiently link attribute and geometry data as in today's GIS were not available at that time. As DBMS technology became available, work on the cadastral register was started earlier, whereas map automation was a byproduct of surveying. Today, however, both systems together form the integrated land information system (LIS), which includes a complex set of procedures and administrative processes for generating and updating land information in a multi-user, multi-agency environment. This system is characterized by a strict separation of the users from the database management system through a standardized interface or data transfer format, the EDBS (Einheitliche Datenbankschnittstelle).

Despite its long and careful planning, implementation of the ALB/ALK system has been slow. North-Rhine Westphalia, the largest Federal state, can serve as an example for this. In the early 1980s some local governments in North-Rhine Westphalia began to digitize cadastral maps. Other local governments started later, and some have not even started today. The ALK has been completed in only two municipalities. It is estimated that the state-wide completion of the digital map system will take another 20 years. Local governments have therefore been advised to confine digitization in a first round to built-up areas. In the other Länder, the situation is no better.

This delay is particularly regrettable as reliable base maps have never been more important than today. In most cities local sewerage systems have to be renewed if further ground water contamination is to be avoided. The ALK is urgently needed as a base map for the inventory of sewer networks and for plans for their renewal. Also utility companies are eager to use the ALK as a base map for their network inventories, even though they would be using only 12 of its about 600 objects (Lutz 1992). The volume of digital data processed by a typical utility company can be demonstrated by the example of the gas and electricity company of the city of Stuttgart (Mahler 1993). For the documentation of the gas network alone, more than 6000 map sheets were produced and are stored on micro-film punched cards. The company is currently reorganizing these maps in digital form. In the beginning it relied on digital city base maps provided by the city surveying department. However, in recent years, more and more digital information from the ALK such as terrain contours is used.

Other problems of the ALB/ALK system are related to its organization. The integration of the existing land register (ALB) and the digital map (ALK) has yet to be realized. For instance, the ALB contains some empty data fields which are to be filled by data from the ALK. One of these data fields should contain the land use category as contained in the legal land use plan for calculating the capacity for development as the difference between existing and zoned land utilization. However, the land use categories of the land use plan are not available, because they are compiled by local government planning departments, and these are external to the cadastral authorities.

In some areas, especially within large agglomerations, land use is changing very fast, so that the land use codes in both ALB and ALK have to be regularly updated. But there are no procedures for regular updating in the organizational system of the cadastre. Updates are performed only on request, with the effect that the land use information in the cadastre is notoriously out of date. In addition, there is no guarantee that ALB and ALK are updated simultaneously, so it is possible that they become inconsistent if the updating takes place at different times (Felletschin, 1993).

The cadastral departments try to improve the accuracy of old cadastral maps when transforming them to digital form. But that would in most cases require a completely new survey and would be costly and time-consuming. Another procedure consists of digitizing existing maps and using sophisticated homogenization software to adjust the digitized maps to known measurement points taking account of known properties of map objects such as linearity, parallelism or orthogonality. However, experiments showed that this is also very time-consuming. Therefore, increasingly the conversion of analogue maps to digital form is performed without improvement in accuracy, either by manual digitizing with high precision digitizers or by scanning and subsequent vectorization. For a later stage, incremental re-surveying of such maps is intended.

ATKIS: A FEDERAL MAP STANDARD

The ALK system is supposed to be the basic GIS standard not only for the local cadastre but also for the topographical maps of the state surveying agencies in order to avoid duplication of digitization by private firms or public agencies and to guarantee the highest map quality. In the late 1980s therefore the official topographic–cartographic information system ATKIS (Amtliches topographisch–kartographisches Informationssystem) was launched as a common initiative of several Länder surveying departments. In the meantime all Länder have joined the project and started with digitizing their analogue maps. North-Rhine Westphalia is one of the leaders in the development of ATKIS and has promised to have ATKIS maps available for the whole state no later than 1996. However, it is not likely that the ATKIS map system will be completed nation-wide before the end of the decade.

ATKIS builds on the experience of the ALK project. The data model is similar to that of the ALK and follows the same object model. It covers all topographic maps of the scales 1:25 000 to 1:100 000. Implementation of ATKIS is performed in two steps. At first topographic objects are digitized only with their geometric features and a limited set of attributes. After the first step, these base map data can be transferred to users via the EDBS. In the second step cartographic features are added to the objects according to their stored attributes.

MERKIS: A SPATIAL REFERENCE SYSTEM

With growing availability of off-the-shelf, low-cost geographical information systems, it is increasingly becoming attractive for local government departments to set up their own GIS independently of other agencies. However, such isolated efforts carry the risk of duplication of data capture and data storage, inconsistency of the spatial reference system and incompatibility of base map geometries.

The Association of German Cities (Deutscher Städtetag) therefore proposed an organizational structure which regulates the responsibility for the provision, maintenance and use of geographic data within local governments called MERKIS (Maßstabsorientierte einheitliche Raumbezugsbasis für kommunale Informationssysteme). MERKIS is a scale-oriented unified spatial reference system for local government information systems. Following MERKIS (Deutscher Städtetag, 1988), each type of geographic base data should be maintained and updated by only one authority and shared by all user agencies. The core principle of MERKIS is that all local government GIS in Germany are to be based on a unique spatial reference system, irrespective of whether or not they are already digitized. In each municipality, only one agency is to be responsible for the administration and updating of the spatial reference system.

The MERKIS recommendation does not suggest a data model nor definitions of objects nor software. It is an organizational structure which:

- defines the technological and organizational requirements for the design and implementation of a unified spatial reference system,

- takes account of existing technological and institutional requirements and existing information systems, and

- determines the conditions under which external users can use the geographical information contained in the systems.

MERKIS defines spatial reference levels (Raumbezugsebenen or RBE), which are to be applied to the tasks of the user agencies according to their needs. Whereas the geographic base data are maintained by a central agency, the user agencies remain responsible for updating their own data. In this way the responsibility for all data in the system is clearly defined. The same applies to the rules by which data in the networked system are exchanged and shared. The EDBS is suggested as the standard data interface for all communication with other government agencies or private users.

The implementation of the spatial levels can be performed independently of each other. Figure 5.1 (Deutscher Städtetag 1988) shows that RBE 500 (Scale 1:500) is the city base map of the ALK. As scale 1:500 is mandatory for the local building plan (Bebauungsplan), the more detailed plan of the two-level German statutory land use planning system, RBE 500 maps are to be digitized first. However, as was discussed above, the slow implementation of the ALK makes this infeasible. Therefore here, too, time-saving digitization or scanning of analogue maps is recommended.

The base map for the less detailed land use plan (Flächennutzungsplan) of the two-level land use planning system is RBE 5000 (Scale 1:5000), the topographic map.

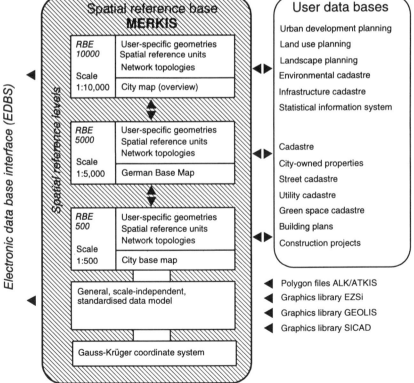

Figure 5.1 MERKIS components and levels. (Source: Deutscher Städtetag 1988).

The derivation of RBE 5000 maps from RBE 500 maps by generalization will only be possible after all RBE 500 maps have been digitized. Therefore RBE 5000 maps will remain analogue or digitized or scanned from analogue maps for some time to come. The same applies to RBE 10 000 or 1:10 000 maps.

At this point the MERKIS recommendation makes a reference to the ATKIS project. Given the intended accuracy of the ATKIS base geometry, it seems feasible to utilize ATKIS objects for the generation of RBE 5000 and RBE 10000 maps (Mittelstraß, 1993). MERKIS is to facilitate overlay and intersection of cartographic objects in spatial, administrative and topological dimension. In addition, MERKIS is to support existing and new GIS applications such as (see Figure 5.2):

- local base maps (city base maps 1:500 to 1:2500, cadastral maps, and German Base Maps 1:2500 to 1:10 000);

- land use plans (building plans, land use plans and landscape plans showing building regulations and spatial policies);

- thematic maps (showing functional relationships and spatial distributions of spatial attributes);

- maps for engineering and landscaping projects and facility management (utility and transport networks);

- maps for districting and location/allocation for public facilities planning (education, health care or sport facilities);

- land use inventory maps (showing existing land use, soil contamination, trees and natural monuments, etc.).

To support a wide range of special applications, MERKIS defines requirements for the geometric component of the data base. It envisages a user-independent data model, in which both universal geometry data (e.g. the base maps for all users) and user-specific geometry data (e.g. the perimeter of a contaminated area) are stored. According to MERKIS, it should be possible to link universal geometries with each other and with user-specific data. Similarly it should be possible to associate attributes, objects and aggregates of objects to predefined or user-defined areal units. At the same time the data model should permit the storage of point, line and polygon objects with as little redundancy as possible.

It has been recognized that in most cases the traditional structure of local government is not adequate for the kind of information management envisaged by MERKIS. Some of the failures of earlier GIS projects can be traced to the neglect of institutional constraints (Cummerwie, 1993).

A central objective of MERKIS is the integration of local government data bases and the avoidance of inefficient duplication of data capture and maintenance. To achieve this, the existing organizational structures within local government need to be modified. This modification is a major intervention into a highly complex organizational system, in which responsibilities, interdependencies and procedures have evolved over a long time. This will be demonstrated using the introduction of MERKIS in the local government of Wuppertal as an example (Cummerwie, 1993).

To introduce MERKIS into the local government of Wuppertal, major reorganizations are necessary, which need to be endorsed by the top of the administration and the city council. For the implementation of the reorganization, a strong organizational management taking account of the decentralized competence and

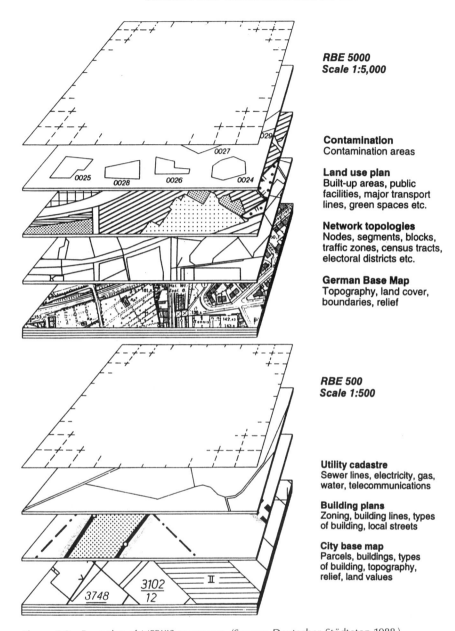

RBE 5000
Scale 1:5,000

Contamination
Contamination areas

Land use plan
Built-up areas, public
facilities, major transport
lines, green spaces etc.

Network topologies
Nodes, segments, blocks,
traffic zones, census tracts,
electoral districts etc.

German Base Map
Topography, land cover,
boundaries, relief

RBE 500
Scale 1:500

Utility cadastre
Sewer lines, electricity, gas,
water, telecommunications

Building plans
Zoning, building lines, types
of building, local streets

City base map
Parcels, buildings, types
of building, topography,
relief, land values

Figure 5.2   Examples of MERKIS coverages. (Source: Deutscher Städtetag 1988.)

responsibility of the user agencies is necessary to make sure that the changes to be made are accepted by all participants. On the other hand, the organizational management has to follow the MERKIS recommendation to install a common spatial data basis. The networking of different information systems requires the application of new interdisciplinary project organizations besides existing hierarchical structures.

The reorganization of well-established organizational structures can only be successful if it is accepted by all participants. Management seminars for leading administrators, workshops and presentations of pilot projects are planned as introduction strategies and are designed to stimulate the interest of involved agencies and departments.

A steering committee was set up to define planning goals and to determine priorities of information processing. In addition, an interdisciplinary working group acts as a mediator between different interests with respect to development, investment, implementation of individual system components, data capture, storage and analysis of data, as well as standardization, data protection and data usage, and training of staff. Based on the goals of the steering committee, the working group prepares decisions, and defines the terms of reference of individual projects. The user agencies are represented by specially trained persons. The implementation of new applications of MERKIS is too complex to be performed in parallel with daily work and therefore will be assigned to project groups released from everyday tasks (Wieser, 1990).

In local governments in Germany, responsibilities are allocated to departments following the recommendations of the KGSt (Kommunale Gemeinschaftsstelle für Verwaltungsvereinfachung), the Local Government Standardization Organization. In Wuppertal, the KGSt organization model will be preserved as much as possible. However, within this framework, the responsibility for the technical infrastructure, organization, user service, data protection, and information management will remain with the central administration department (Hauptamt). The statistics department will be in charge of making available spatial and statistical data and codes and of maintaining the address reference system. The surveying department will be responsible for linking geometric with attribute data, computer graphics and the maintenance of the graphical database. Responsibility for the user data, however, including system supervision, user-specific geometries, statistical analyses, simulation and forecasts will rest with the user agencies.

Only recently, the KGSt has started encouraging cities to follow the MERKIS recommendation (Kommunale Verwaltungsstelle für Verwaltungsvereinfachung, 1994). Currently some ten other large cities besides Wuppertal have begun to reorganize their adminstration in the spirit of MERKIS.

A GIS SURVEY

The most recent surveys of computer applications in local government in Germany date back to the late 1970s and early 1980s, i.e. to pre-GIS times (Wegener, 1983). In 1993, the Commission of the European Communities (DG XIII) conducted a survey of the diffusion of GIS in medium-sized cities (with populations between 100 000 and 400 000) in France, Germany, Greece and Spain. In Germany the DIFU (German Institute of Urbanism), a research institute of the Deutscher Städtetag, distributed the EC questionnaire and received usable answers from 40 cities. In addition, 15 cities were surveyed with a more detailed questionnaire. The results of the four countries will be assembled as a handbook.

The EC study coincided with the GISDATA survey of GIS diffusion in local governments in Europe. In the German part of the GISDATA survey, 86 cities with a population of more than 100 000 were asked about their GIS experience and plans following the methodology developed at the University of Sheffield. The survey was conducted in close cooperation with the Deutscher Städtetag and the KGSt. Both institutions expressed a strong interest in the GISDATA study and suggested adding a few questions of interest to them to the GISDATA questionnaire, such as questions regarding the organization of GIS, the cooperation between user-agencies and the access of external users to GIS data. The cooperation with the Deutscher Städtetag and the KGSt was an important door opener for the survey.

Of the 86 cities, 80 returned the completed questionnaire. Of these, four are in the largest size category with more than 700 000 population, 11 in the second with population between 400 000 and 700 000, 22 in the third with population between 200 000 and 400 000 and 49 in the fourth category with population between 100 000 and 200 000 inhabitants. Only one city refused to answer because of lack of time to complete the questionnaire. Figure 5.3 shows the location of the 86 cities by size category. It can be seen that a high proportion of cities is in the largest Federal state, North-Rhine Westphalia, whereas the new Länder have only few large cities.

It may be of interest to note which local government department returned the questionnaire. In all cases the questionnaire was addressed to the Hauptamt which returned it completed in 40% of the cases. However, 46% of the questionnaires were completed by the surveying department. This confirms the dominant role of the surveying discipline in the introduction of GIS in Germany (Wegener and Junius, 1993)—a fact that will become even more apparent in the analysis of the survey findings.

Diffusion of GIS

The first surprising result of the survey is that of the 80 cities that completed the questionnaire 70 indicated that they are using some kind of GIS or GIS predecessor or component, and that the remaining ten cities stated that they are planning to purchase a GIS either within a year (seven cities) or in the near future (three cities). Table 5.1 shows the distribution of GIS diffusion across city size categories. It can be seen that with one exception the late adopters belong to the smallest size group of cities in the survey.

Table 5.1 Diffusion of GIS in German cities by city size, 1994

City size (× 1000)		No response	GIS component planned	GIS component in operation	Sub-total	Total
100–200	No.	3	9	37	46	49
	%	6.1	18.4	75.5	93.9	100.0
200–400	No.	3	1	18	19	22
	%	13.6	4.5	81.8	86.4	100.0
400–700	No.	0	0	11	11	11
	%	0.0	0.0	100.0	100.0	100.0
over 700	No.	0	0	4	4	4
	%	0.0	0.0	100.0	100.0	100.0
Total		6	10	70	80	86
	%	7.0	11.6	81.4	93.0	100.0

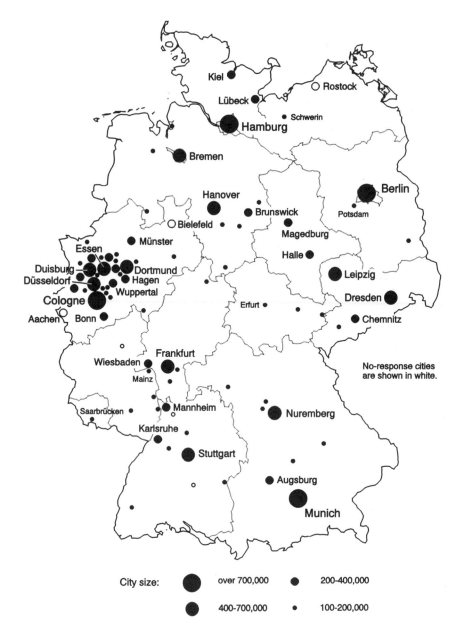

Figure 5.3 Location of the 86 cities surveyed by population size category.

This result seems to indicate that the diffusion of GIS in large German cities has almost been completed. However, a look at Figure 5.4 suggests a more cautious interpretation. Figure 5.4 shows the penetration of GIS and GIS components in the 70 cities over time based on cumulated purchases of software systems. It can be seen that only 45 of the 70 cities possess a full-scale integrated geographic information system superseding, or possibly complementing, earlier less integrated components for data capture (digitizing), data management or data analysis and presentation.

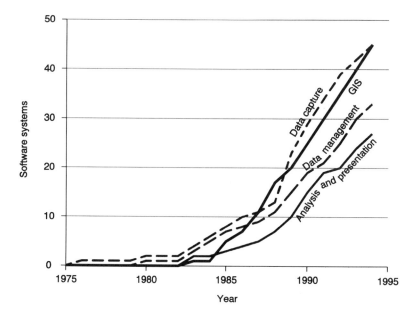

Figure 5.4 Diffusion of GIS in German cities 1975–1994.

The figure clearly conveys that the S-shaped GIS diffusion curve is only in its mid-phase, with saturation effects still some time ahead.

Table 5.2 shows the software used. Altogether 25 software systems were mentioned. There are two clear leaders, SICAD by Siemens-Nixdorf and ALK-GIAP, the custom-tailored surveying and mapping software developed by AED, Bonn, for the ALK working group. The success of the latter is partly explained by the fact that cities can get the ALK-GIAP software practically free from the cooperating Länder agencies. Both SICAD and ALK-GIAP mainly address the needs of surveying

Table 5.2 GIS software in German cities, 1994

GIS software	Number of systems
SICAD (Siemens-Nixdorf)	31 (34.1%)
ALK-GIAP (AED)	17 (18.7%)
GEOLIS/GEIS (IBM)	7 (7.7%)
EZS-Interaktiv (Mannesmann-Kienzle)	5 (5.5%)
ARC/INFO (E.S.R.I.)	4 (4.4%)
MGE (Intergraph)	3 (3.3%)
PROCART (CONDATA)	3 (3.3%)
ATLAS-GIS (Strategic Mapping Inc.)	2 (2.2%)
AutoCAD (Autodesk)	2 (2.2%)
DAVID (Ingenieurbüro Riemer)	2 (2.2%)
Others[a]	15 (16.5%)
Total	91 (100.0%)

[a] DCS, DIGSY, GEO-GPG, Geograph, GRADIS, GRIPS, GTI-RDB, NPK, PCmap, Planum, Smallworld, SPANS, TRIAS, winCAD, YADE.

departments and so again reflect the importance of surveying for the diffusion of GIS in German local governments. None of the other software packages is mentioned more than five times, and 15 only once; some are not full GIS but digitizing or mapping software. Table 5.2 does not fit well together with Figure 5.4; whereas Figure 5.4 plots a total of 150 purchases, Table 5.2 lists only 91 software systems. The reason is that the respondents only in part entered their GIS predecessors, i.e. digitizing, mapping or data analysis, systems.

GIS Applications

In order to find out for which tasks GIS are used in German cities, 16 applications in which GIS are typically applied were selected for the analysis. For the presentation in this chapter, the 16 applications were aggregated into five application fields, surveying, statistics, utilities, planning, and others as indicated in Table 5.3.

It was asked in the questionnaire in which application field GIS were used or planned to be used in the near future. This information was then compared with the maximum number of potential applications, i.e. the number of applications which would result if in each city each possible GIS application would have been implemented—for 70 cities and 16 possible applications this yields 70 times 16 or 1120 potential applications.

Table 5.4 shows the GIS applications in the 70 cities aggregated for the five application fields. It needs to be pointed out that the absolute number of applications in each field reflects the level of its disaggregation into applications but not its magnitude or importance. For instance, the application field surveying consists of only one relatively homogeneous but very large application, whereas the application field planning was broken down into six relatively small different applications. Therefore not the absolute numbers but their percentages are indicative of the rate

Table 5.3 GIS application fields

Application field	Application
Surveying	Surveying and cadastre
Statistics	Statistics
Utilities	Water, gas, electricity
	Sewerage
	Waste disposal
Planning	Urban development
	Land use planning
	Building plans (Bebauungsplanung)
	Landscape planning, green space ordinances
	Environmental planning/protection
	Transport planning
Others	City-owned properties
	Traffic (networks, traffic management) Building permits
	Social infrastructure
	Emergency services

Table 5.4 GIS applications in German cities by application field, 1994

Application field		No information	No application	Actual applications				Potential applications
				Concept phase	Pilot phase	Daily use	Sub-total	
Surveying	No.	0	0	4	12	54	70	70
	%	0.0	0.0	5.7	17.1	77.1	100.0	100.0
Statistics	No.	7	19	8	12	24	44	70
	%	10.0	27.1	11.4	17.1	34.3	62.9	100.0
Utilities	No.	37	73	32	26	42	100	210
	%	17.6	34.8	15.2	12.4	20.0	47.6	100.0
Planning	No.	49	123	109	86	53	248	420
	%	11.7	29.3	26.0	20.5	12.6	59.0	100.0
Others	No.	64	177	57	17	35	109	350
	%	18.3	50.6	16.3	4.9	10.0	31.1	100.0
Total		157	392	210	153	208	571	1120
	%	14.0	35.0	18.8	13.7	18.6	51.0	100.0

of diffusion of GIS in the application fields. It can be seen that the diffusion of GIS has indeed progressed fastest in surveying. In more than three quarters of all cities surveying departments routinely apply GIS technology in their daily work, in the remaining cities GIS application for surveying is in a planning or testing stage; there is not a single city in which surveying is without GIS support. All other application fields are far behind. Statistical offices are second, with about one third of potential GIS applications implemented; utilities third with one fifth, and planning fourth with about 13% of potential applications. Altogether only 19% of all potential GIS applications in large German cities have been realized.

Table 5.5 shows the same information aggregated by city size. Surprisingly not

Table 5.5 GIS applications in German cities by city size, 1994

City size (× 1000)		No information	No application	Actual applications				Potential applications
				Concept phase	Pilot phase	Daily use	Sub-total	
100–200	No.	98	226	92	67	93	252	592
	%	16.6	38.9	17.1	11.3	16.2	44.6	100.0
200–400	No.	36	98	54	40	60	154	288
	%	12.5	34.0	18.8	13.9	20.8	53.5	100.0
400–700	No.	23	40	42	33	38	113	176
	%	13.1	22.7	23.9	18.8	21.6	64.2	100.0
over 700	No.	0	24	13	13	14	40	64
	%	0.0	37.5	20.3	20.3	21.9	62.5	100.0
Total	No.	157	392	210	153	208	571	1120
	%	14.0	35.0	18.8	13.7	18.6	51.0	100.0

the largest cities but cities in the upper middle range, between 400 000 and 700 000 population, are the most active adopters of GIS; they report two out of three potential GIS applications either in daily use or in the planning and testing phases (though the relative large percentage of 'no-application' entries in the largest size category suggests that there may be some underreporting there). As expected, implementation of GIS applications in smaller cities is lagging behind.

Spatial Data Bases

A separate question addresses the state of data capture for the above applications. Table 5.6 presents the implementation of spatial data bases in the 70 cities by application field. Again not the absolute numbers but the percentage of actual versus potential implementations is of interest. Here, too, surveying leads the field, with spatial data bases for base maps and cadastre in all cities either completed or under development, although data capture is still in progress in three out of four cities, even in cities in which a GIS is already in daily use (see Table 5.4). Next to surveying, statistical departments report a relatively high percentage of completed spatial data bases; the reason for this may be that in this case only a spatial reference system and a base map had to be added to an otherwise already computerized data base. Planning is, as usual, lagging behind, although (not shown) two out of three cities have digitized their land use plans or intend to do so in the near future. Altogether, spatial data bases have been completed or are under development for less than half of all potential local government applications; so data capture is likely to remain one of the major tasks of GIS diffusion in the future. A comparison between Tables 5.4 and 5.6 shows that data capture is lagging behind application development. There are many GIS applications already in daily use in which data capture is still in progress, and many applications are being conceptualized for which no spatial data base yet exists.

Table 5.6 Diffusion of GIS in German cities by application field, 1994

Application field		No information	No application	Actual applications				Potential applications
				Concept phase	Pilot phase	Daily use	Sub-total	
Surveying	No.	0	0	5	52	13	70	70
	%	0.0	0.0	7.1	74.3	18.6	100.0	100.0
Statistics	No.	10	20	9	20	11	40	70
	%	14.3	28.6	12.9	28.6	15.7	57.1	100.0
Utilities	No.	45	70	25	58	12	95	210
	%	21.4	33.3	11.9	27.6	5.7	45.2	100.0
Planning	No.	66	144	74	107	29	210	420
	%	15.7	34.3	17.6	25.5	6.9	50.0	100.0
Others	No.	78	166	41	45	20	106	350
	%	22.3	47.4	11.7	12.9	5.7	30.3	100.0
Total	No.	199	400	154	282	85	521	1120
	%	17.8	35.7	13.8	25.2	7.6	46.5	100.0

Table 5.7 Spatial data bases in German cities by application field, 1994

Application field		Department specific	ALK	German Base Map 1:5000	ATKIS	Total
Surveying	No.	19	56	7	2	84
	%	22.6	66.7	8.3	2.4	100.0
Statistics	No.	20	9	14	4	47
	%	42.6	19.1	29.8	8.5	100.0
Utilities	No.	29	53	9	6	97
	%	29.9	54.6	9.3	6.2	100.0
Planning	No.	77	77	53	22	229
	%	33.6	33.6	23.1	9.6	100.0
Others	No.	33	41	15	10	99
	%	33.3	41.4	15.2	10.1	100.0
Total	No.	178	236	98	44	556
	%	32.0	42.4	17.6	7.9	100.0

Next it was asked which spatial reference systems were adopted for the spatial data bases. Table 5.7 shows this, again by application field. It is clear that the parcel-based ALK spatial reference system is dominant, in particular in surveying and utilities, witnessing a successful standardization between departments and between cities. For smaller-scale maps in statistical analysis and planning, spatial reference systems based on the German Base Map 1:5000 are used, even though in both application fields department-specific spatial reference systems are also frequent. The ATKIS map system is still rarely used, presumably because of its delayed implementation. There are more spatial reference systems than spatial data bases; in some application fields, as in surveying, some cities use more than one spatial reference system.

How many of the spatial reference systems follow the MERKIS recommendations? Table 5.8 shows that the basic spatial reference level, RBE 500 (Scale 1:500), has the highest implementation rate, followed by RBE 5000 (Scale 1:5000) and RBE 10000 (Scale 1:10000). There are almost no implementations at the remaining map scales. Altogether the MERKIS recommendation is followed by only one out of three potential applications. This indicates that considerable efforts to replace department-specific spatial reference systems and to integrate spatial reference levels and spatial data bases of different departments and agencies are still needed.

Hardware

Table 5.9 shows which type of hardware is used for GIS activities in German cities. There is still a relatively large proportion of mainframe applications (because until recently SICAD and GEOLIS/GEIS required mainframes), however workstations and high-end personal computers are rapidly catching up and are likely to largely displace mainframes soon. In particular the high rate of adoption of Unix-based

Table 5.8   Adoption of MERKIS recommendations in German cities, 1994

MERKIS level	Scale		No information	Adoption of MERKIS Planned	Completed	Sub-total	Potential MERKIS implementation
RBE 500	1:500	No.	8	49	13	62	70
		%	11.4	70.0	18.6	88.6	100.0
	1:1000	No.	69	1	0	1	70
		%	98.6	1.4	0.0	1.4	100.0
	1:2500	No.	68	0	2	2	70
		%	97.1	0.0	2.9	2.9	100.0
RBE 5000	1:5000	No.	25	23	22	45	70
		%	35.7	32.9	31.4	64.9	100.0
RBE 10 000	1:10 000	No.	35	25	10	35	70
		%	50.0	35.7	14.3	50.0	100.0
	1:20 000	No.	68	0	2	2	70
		%	97.1	0.0	2.9	2.9	100.0
Total			273	98	49	147	420
		%	65.0	23.3	11.7	35.0	100.0

Table 5.9   Hardware used in German cities by application field, 1994

Application field		Mainframe	Work-station	PC	Multi-platform	Multi-platform network	Total
Surveying	No.	11	18	3	12	26	70
	%	15.7	25.7	4.3	17.1	37.1	100.0
Statistics	No.	7	8	9	7	10	41
	%	17.1	19.5	22.0	17.1	24.4	100.0
Utilities	No.	20	18	17	19	17	91
	%	22.0	19.8	18.7	20.9	18.7	100.0
Planning	No.	25	64	44	20	43	196
	%	12.8	32.7	22.4	10.2	21.9	100.0
Others	No.	28	15	17	8	28	96
	%	29.2	15.6	17.7	8.3	29.2	100.0
Total		91	123	90	66	124	494
	%	18.4	24.9	18.2	13.3	25.1	100.0

ALK-GIAP will promote the use of workstations. Multiplatform stand-alone and network solutions are gaining in importance.

First Experience

As GIS in local government are a relatively recent phenomenon (see Figure 5.4), it may be too early to assess the benefits resulting from their introduction.

Table 5.10 Most important perceived benefits of GIS, 1994

		No information	Magnitude of benefit			Total
			Large (Rank 1)	Medium (Rank 2)	Small (Rank 3)	
Kind of benefit						
Better information	No.	2	45	15	8	70
	%	2.9	64.3	21.4	11.4	100.0
Better decisions	No.	4	21	33	12	70
	%	5.7	30.0	47.1	17.1	100.0
Savings	No.	11	8	14	37	70
	%	15.7	11.4	20.0	52.9	100.0
Total		17	74	62	57	210
	%	8.1	35.2	29.5	27.1	100.0

Nevertheless there is a clear ranking of these impacts by the responding cities (see Table 5.10). A total of 45 cities stated that the most important benefit of GIS has been that they have improved information processing. Faster information retrieval and enhanced possibilities for data analysis and presentation were most frequently cited as reasons for this. Only 21 cities held the opinion that the most important possible impact of GIS has been that better information has also improved decision making, though mostly in the field of operational, as opposed to strategic or managerial, decisions. Only a small minority thought that the most important effect of the introduction of GIS relates to savings. These are savings in time, if any, but hardly ever in staff or cost.

There are fewer entries on problems encountered during the introduction of GIS than on perceived benefits, which may indicate that benefits have outweighed problems. The largest number of respondents, 27, stated that organizational problems have been the most severe (Table 5.11). Of these, lack of qualified staff was quoted most frequently, followed by insufficient motivation of staff by management. Only

Table 5.11 Most important problems in the introduction of GIS in German cities, 1994

		No information	Severity of problem			Total
			Severe (Rank 1)	Medium (Rank 2)	Small (Rank 3)	
Kind of problem						
Organizational problems	No.	21	27	9	13	70
	%	30.0	38.6	12.9	18.6	100.0
Technical problems	No.	22	14	25	9	70
	%	31.4	20.0	35.7	12.9	100.0
Data problems	No.	25	13	14	18	70
	%	35.7	18.6	20.0	25.7	100.0
Total		68	54	48	40	210
	%	32.4	25.7	22.9	19.0	100.0

half as many respondents thought that technical or data problems were most important, though more respondents ranked technical problems such as lack of user-friendliness or lack of compatibility of GIS software, in the first place. Data problems were mainly related to the cost of data capture and updating.

CONCLUSIONS

The presentation of the GIS scene in local governments in Germany shows the special situation in a country in which the diffusion of GIS is influenced by the federal organization of government and, within local governments, by surveying and cadastral departments. This combination has led to a slow diffusion of GIS compared with other countries in Europe.

The survey confirms the delay in GIS diffusion in local governments in Germany. Of the 80 cities responding to the survey, 70 have some experience with GIS components, but only 45 have to date acquired full-scale integrated GIS. At the level of GIS applications the gap between actual and potential GIS use is even larger: less than one fifth of all potential GIS applications have been implemented or are in the planning stage. Surprisingly, not the largest cities but the cities in the upper mid-range of city sizes are the most active in adopting GIS technology. However, where GIS applications are being implemented, spatial data bases and spatial reference systems tend to be well integrated between different map scales and departments and agencies. As all cities presently without a GIS or GIS components stated their intention to adopt GIS technology, it can be expected that GIS diffusion in German cities will make rapid progress in the near future.

The initially slow adoption of GIS technology in German cities has made it possible to pay more attention to the organisational integration of GIS into the existing institutional framework and to problems of data integration and cooperation between different departments in local government, between different levels of government and between local governments and other public and private GIS users. In particular the MERKIS recommendation, which is now being taken up by a growing number of cities, seems to provide a basis to avoid mistakes made elsewhere and arrive at a well organized and integrated system. It remains to be seen whether the German example of carefully planned adoption of the new technology is worth the delay in adoption, or whether the speed of technological advance in the GIS field will render all planning efforts futile.

ACKNOWLEDGEMENTS

The authors are grateful to Oliver Rathjens and Stephan Wilforth who assisted in pre-processing and analysing the results of the survey.

REFERENCES

CUMMERWIE, H.-G., 1993. Organisationsstrategie für den Aufbau kommunaler GIS und Realisierung von MERKIS in Wuppertal (Organizational strategy for the development of local government GIS and the implementation of MERKIS in Wuppertal). Mimeo. Surveying Department, City of Wuppertal.

DEUTSCHER STÄDTETAG, 1988. Maßstabsorientierte Einheitliche Raumbezugsbasis für Kommunale Informationssysteme (MERKIS) (Scale-oriented unified spatial reference base for local government information systems). DST-Beiträge zur Stadtentwicklung und zum Umweltschutz, Reihe E. Deutscher Städtetag, Cologne.

FELLETSCHIN, V., 1993. Beitrag des Vermessungswesens zum Aufbau raumbezogener Informationssysteme (Contribution of surveying to the implementation of spatial information systems). *Vermessungswesen und Raumordnung*, **2**, pp. 92–100.

KOMMUNALE GEMEINSCHAFTSSTELLE FÜR VERWALTUNGSVEREINFACHUNG, 1994. Raumbezogene Informationsverarbeitung in Kommunalverwaltungen (Spatial information processing in local governments). *KGSt-Bericht*, **12**, KGSt, Cologne.

LUTZ, W., 1992. Anlagen der Energieversorgung eingebunden in ein Geo-Informationssystem (Power generation plants integrated into a geographic information system). *Geoinformationssysteme*, **5**, 7–15.

MAHLER, D., 1993. Anforderungen eines Leitungsbetreibers an die Graphische Datenverarbeitung (Requirements of a utility company with respect to computer graphics). *Vermessungswesen und Raumordnung*, **2**, 63–74.

MITTELSTRAß, G., 1993. Verbindung zwischen ALK, ATKIS und MERKIS (Link between ALK, ATKIS and MERKIS). *Zeitschrift für Vermessungswesen*, **5**, 242–8.

WEGENER, M., 1983. The impact of systems analysis on urban planning: the west German experience, in Hutchinson, G.B. and Batty, M. (Eds) *Systems Analysis in Urban Policy Making and Planning*, pp. 125–52, New York: Plenum.

WEGENER, M. and JUNIUS, H., 1991. 'Universal' GIS versus national land information systems: internationalization of software or endogenous development?, *Computers, Environment and Urban Systems*, **15**(4), 219–27.

WEGENER, M. and JUNIUS, H., 1993. 'Universal' GIS versus national land information traditions: software imperialism or endogenous developments?, in Masser, I. and Onsrud, H.J. (Eds) *Diffusion and Use of Geographic Information Technologies*, pp. 213–228, Dordrecht: Kluwer Academic Publishers.

WIESER, E., 1990. Projektmanagement. *Allgemeine Vermessungsnachrichten* **11–12**, 397–402.

CHAPTER 6

Italy: GIS and administrative decentralization

LUISA CIANCARELLA, MASSIMO CRAGLIA, ENZO RAVAGLIA, PIERO SECONDINI and EDI VALPREDA

INTRODUCTION

This chapter reviews the diffusion of GIS in Italian local government based on two comprehensive surveys of Regions and Provinces carried out in 1993–94 by the Environment Department of ENEA, the Italian agency for New Technologies, Energy and the Environment, in partnership with the Universities of Sheffield and Bologna. In the absence of a comprehensive survey among the over 8000 municipalities in Italy, information on GIS diffusion at this level is based on a number of studies carried out among metropolitan areas (Jogan and Schiavoni, 1993; Ciancarella *et al.*, 1992), and medium-sized communes (Craglia, 1993).

The overall findings of the surveys and studies presented in this chapter indicate an increasing awareness of the potential of GIS and growing take-up of this technology in Italian local government. One of the main contributing factors for this greater attention towards GIS appears to be the 1990 reform of local government which increases the decentralization of power from central and regional government towards Provinces and Municipalities.

To give context to the research findings, the following sections provide an overview of information technology in local government, availability of digital topographic data and structure of Italian local government. A summary of recent political events is also included as these are likely to have profound effects on the Italian system of government, and hence on the institutional context within which GIS diffusion is analysed.

NATIONAL CONTEXT

Italy is presently undergoing major political and social change. Since 1992, the political parties that governed Italy since World War II have all but disappeared, a new electoral system has been introduced, and an entirely new political coalition has won the 1994 general elections. This coalition only lasted until January 1995 and new general and local elections are taking place in 1995. While the political scene is changing rapidly and is characterized by a great deal of uncertainty and instability, there is a widespread view that no significant improvement of local governance has taken place yet.

The studies analysed in this chapter have been carried out during this period of transition between the 'old' system and a largely unknown 'new' one, and reflect the uncertainty felt throughout the system of public administration. Change and instability create both potential for and constraints to the diffusion of technology. There is ample evidence that introducing a new technology, like GIS, to an organization involves changes in individual and organizational practices and challenges existing power relationships (Barrett, 1992; Craglia 1993). The insecurity and anxiety thus created may be more difficult to manage when the environment within which the organization is embedded is also undergoing major change. Moreover, the long lead time required to implement GIS suggests that by the time the system is operational it may have to meet a completely new configuration of policies and programmes from that for which it was designed, thus creating the conditions for a less than effective utilization. For this reason, Masser and Campbell (1991) argued that organizational and environmental stability is critical to the effective utilization of information technology in local authorities. In this sense the current developments in Italy and the high level of instability they generate in public administration may act as constraints to the diffusion and effective utilization of GIS in local government. On the other hand, the evidence presented in this chapter indicates that political and institutional change may also provide new opportunities for innovating.

Information Technology (IT) in Local Government

Given the comparative nature of this book on the diffusion of GIS in local government in Europe, it is important to bear in mind that there are national differences in the degree of familiarity of local authorities with computing, which affects their awareness of the possibilities of the technology. As argued by the Chorley Commission in the UK (DOE, 1987), lack of awareness is an important barrier in the take-up of a technology like GIS.

In Italy, the widespread diffusion of IT in local government is only a recent phenomenon by comparison with other European and OECD countries. For example, already in the mid-1970s computing was available in almost all Danish municipalities and in well over half of those in Sweden, the UK, Japan, and the USA. By contrast only 3% of the municipalities with over 2000 people had access to computing in Italy, and only 10 years later, in the mid-1980s, the majority of large Italian municipalities had access to computing. There are two main factors that have contributed to this late start. The first is the larger number of authorities and their smaller average population size in Italy compared to the countries of Northern Europe. Clearly, very small authorities do not 'need' computing and often do not have the human resources to support it (financial considerations were also relevant in the past, much less now). The relationship between population size and availability of computing is clearly illustrated by the findings that 80% of larger Italian authorities (over 100 000 population, representing 1% of all municipalities) had access to computing in 1983 while smaller authorities were far behind (Limone, 1989).

Table 6.1 IT expenditure (1992 US $) in selected OECD countries, 1992

Country	France	Germany	UK	Italy
IT market ($ Mil.)	25 082	33 900	22 173	15 399
IT expenditure/GDP	1.86%	1.92%	2.09%	1.24%

Source: IDC, 1994.

The second, and crucial factor contributing to the delay in the diffusion of computing in Italy relates to the low expenditure on IT. Despite being the fifth largest industrialized nation, and the fourth largest European market for IT, Italy consistently spends some 40% less on IT than its main European competitors (France, Germany, UK). For example Table 6.1 shows that in 1992 Italy spent 1.2% of its Gross Domestic Product (GDP) on IT compared to 1.9%–2.1% in the other three countries considered. Similar findings were reported for 1987 by an Italian research institute (Censis, 1990): Italy 1.4%, France and Germany 1.9%. UK 2.3%. The relatively lower investments on IT by Italy as a whole (including the private sector) cause concern because they make catching up with the main economic competitors more difficult.

Local government, which accounts for approximately 5% of the total IT expenditure in Italy, has been affected by these relatively lower investments particularly in recent years which have been characterized by political instability and severe budget cuts. Despite these constraints, there are indications of a growing awareness in local government of the strategic importance of IT to increase efficiency and improve the level of services. Computer use is becoming more widespread and is being extended from the basic automation of administrative processes to higher level functions. These are the findings of a recent survey by the Department for Public Administration (Dipartimento della Funzione Pubblica, 1993) which show that all local authorities above 50 000 population now have access to computing, and that applications related to land-use planning and environmental monitoring have become the third most common after population registers and finance. An interesting finding of this study which acts as background to the rapid diffusion of GIS among the Italian provinces reported before, is that among the provincial authorities applications on planning and environmental matters are the most common use of computing (15.5% of total computer use). In 1991, GIS expenditure in Italy accounted for some GBP 130 million or 2% of the overall IT expenditure. It was however estimated that in 1991–3 it grew by some 11% in the public sector (with kind permission of Teknibank, 1993, 'GIS Marketing opportunities') suggesting that this market has considerable potential for expansion.

A recent initiative by central government that offers new opportunities for the development of IT in public administration including local government is the setting up in 1993 of a national IT Authority with the objective of developing policies and strategies for IT diffusion and its related organizational and human resources aspects (AIPA, 1994). Particular attention is being given to facilitating the connectivity and information exchange among existing information systems in central and local government, which traditionally have been independent from one another. This is a daunting task, but offers real opportunities to innovate the operational and strategic activities of public administration in general and local government in particular.

Topographic Data

Mapping responsibilities in Italy are fragmented among several institutions. The Italian Geographic Military Institute (IGMI) is responsible for the maps 1:25 000–1:1 000 000 scale. Maps at 1:250 000–1:1 000 000 scale are available for the whole country in raster format while the database associated with the vector format for the 1:200 000 scale map has recently been made available. Efforts are also under way to develop the 1:50 000 scale vector coverage and digitize the topographic features at 1:25 000 scale. These data are available in DIGEST transfer format.

The Cadastral Administration of the Ministry of Finance is the main source for 1:1000–1:4000 scale maps. Since 1986, the Cadastre has launched a major programme to update its maps and records and convert them in digital format. The first completed stage of this programme has been to develop a new national network of control points to support ground field-work. Although the size of this automating and updating task is considerable, progress has been made. By 1993, 85 000 map sheets had been digitized (27% of the total 310 000). Moreover, all new property transactions, which include a local survey, are handled directly in digital form by each of the 93 provincial offices of the Cadastral Administration. To date only 27 Provinces have complete digital base maps including both cadastral and topographic data. To extend the coverage to all other provinces, the Cadastre plans to rasterize updated aerial photographs for some topographic features only, such as road network, hydrography, and limits of the built-up area. Although this initiative is meant to provide a basic support for GIS applications at the municipal level in a relatively short time, the limited precision of the maps thus available raises some questions as to their usefulness for detailed urban planning and management.

The Cadastre has recently started selling its digital maps with special offers to local authorities. It is estimated that the ratio between the cost of these maps and that of acquiring them independently is 1:5 (Cannafoglia, 1993). Clearly, those authorities for which digital Cadastral maps are available may find this a worthwhile opportunity even though cadastral maps are thematic in nature and do not have elevation data. It must be appreciated however, that it will take many more years before all urban areas have digital cadastral maps available and therefore, the current practice of sub-contracting the production of large scale maps to private companies is likely to continue for some time. The maps of the cadastral Administration are made available in a modified version of the British NTF format. However, it would appear that a change towards DIGEST is being considered as this is increasingly becoming the national standard.

Regional authorities are also producing maps to support the land use and infrastructure planning functions devolved to them in the 1970s. The focus is on 1:5000–1:10 000 scale maps (on paper support and more recently in digital format) to fill the large gap left by both IGMI and the Cadastre. Inevitably, each region has tended to produce maps for its own needs at different scales, standards, and level of currency. A measure of the existing variations in paper map availability is that in 1992 there were still no maps at 1:5000–1:10 000 scale for 53% of the country with notable regional variations: 26% of the land area of Northern Italy, 49% in Central Italy, and 82% in the South. Moreover, the average age of the maps available was 14.5 years (Bezoari and Selvini, 1992).

Another new source of digital data is being provided by the national census bureau (ISTAT). For the 1991 census, a programme to digitize the administrative boundaries of the 8000 communes, 50 000 urbanized areas, and 350 000 enumeration districts has been set up at 1:25 000 scale (Orasi and Gargano, 1993). Census data will be geo-referenced to these polygons thus providing a new essential tool for socio-economic analysis in a variety of disciplines and application domains. Given the continuous change in ED boundaries, it will not be possible at this stage to analyse spatial time series of census data at sub-communal level. However, it is one of the aims of ISTAT to use this new digital base and remote sensing technology to update more effectively future changes in the boundaries.

Other national organizations have been producing digital maps to support their operations. These include the national electricity board, the national telephone company, and the railway company. By and large, these efforts have involved digitizing or scanning a variety of paper maps depending on the scale needed (up to 1:200 for the utility companies) and then adding a layer with the company's networks (Ricci and Silvestri, 1993). These efforts have often involved significant investments and have occurred in an uncoordinated fashion and in some cases have been made obsolete by the new developments of IGMI or the Cadastral Administration. On the other hand it is also clear that the lack of national directives and strategy, and the delay of the institutional agencies to produce national digital maps at different scales has forced many organizations to take the initiative.

Structure of Local Government

Italian local government has a three-tier structure comprising of 20 regions, 103 provinces, and some 8100 communes. Table 6.2 shows the average population size and the very broad range of population levels within each tier. At all three levels, local authorities are governed by an executive nominated by an elected assembly. Since 1992, citizens' participation at the communal and provincial level has been strengthened by the direct election of the mayor whose powers in forming the executive have also been increased.

The progressive decentralization of power from central to local government has occurred in two main phases. The 1970s saw the enpowerment of regions to legislate on matters of local economic development, public and social services, infrastructure, natural resources and land-use, and to organize their internal structure and administration. Provinces and communes continued to have minimal autonomy and act as decentralized offices of central government in the fields of infrastructure and education (provinces), local planning, traffic control, food markets and retailing, and the collection of minor local taxes (communes). Local mayors were also responsible

Table 6.2 Population range and average of Italian local authorities, 1991

Level	Population Min.	Max.	Average
Regiona	115 938	8 856 074	2 838 902
Provinces	91 942	3 761 067	551 243
Communes	Approx. 30	2 775 250	7010

Source: ISTAT, 1993

to central government for maintaining the population and electoral registers up-to-date.

The second phase of decentralization has taken place in 1990 with a wide-ranging reform of local government, which changed the overall approach to administration from top-down to bottom-up. The reform (Law 142/90) establishes that local communities, organized in communes and provinces, are autonomous. The commune is identified as the basic level of government, representing its community, and looking after its social and economic well-being, while the province acts as intermediary with the region. This new role as intermediary authority implies that provinces are no longer viewed as decentralized government offices of secondary importance but that they must act as the governing body for the provincial territory in their own right and mediate between the different needs and demands of their communes (Pazzaglia, 1993). The extended responsibilities of the provinces include financial and land-use planning, environmental management, local infrastructure, secondary education and vocational training, public health, data collection and processing, and technical assistance to communes in the field of information technology.

To increase local autonomy, each commune and province is given by this reform the right to draft and adopt an individual charter defining its objectives, setting priorities, and organizing its internal administration and structure accordingly. Moreover, a large degree of financial autonomy (i.e. power to raise local taxes) is envisaged to meet the set objectives.

The reform also established eight new provinces (raising their number from 95 to 103) and 12 new metropolitan authorities to govern the larger cities and their surrounding areas. The exact relationship between these new metropolitan authorities and the overlapping provinces is as yet unclear.

GIS DIFFUSION IN THE REGIONS

Data on the diffusion of GIS among the 20 Italian regions was gathered through a postal survey in 1992 (Ciancarella *et al.*, 1993). That study was completed and updated with a telephone survey carried out between October 1993 and January 1994 which followed the methodological framework developed for the British surveys (Campbell and Masser, 1991). The findings of this new study indicate that 13 regions already have GIS technology (65%) and another two have plans to acquire it in the future. The geographical distribution is shown in Table 6.3 and Figure 6.1. Together they highlight that diffusion started in Northern Italy and is now extending to all

Table 6.3   Plans for GIS by Italian regions, 1994

	Italy		North		Centre		South	
Plans for GIS	No.	%	No.	%	No.	%	No.	%
Already have GIS	13	65	8	100	3	60	2	29
Have plans for GIS sometime in the future	2	10	0	0	1	20	1	14
Have no plans to acquire GIS	5	25	0	0	1	20	4	57
Total	20	100	8	100	5	100	7	100

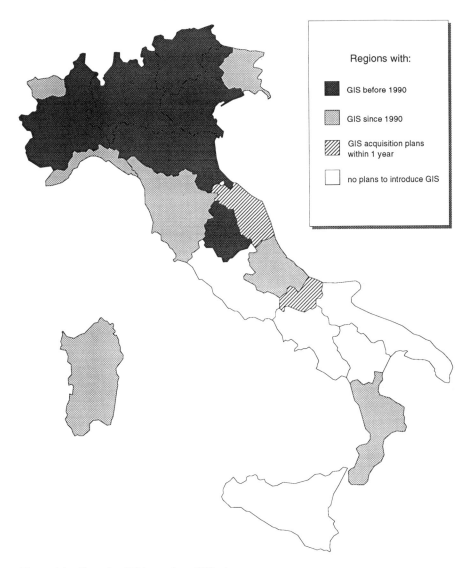

Figure 6.1 Plans for GIS by region, 1993–4.

the Central regions, with the noticeable exception of Lazio. In Southern Italy the diffusion of GIS is as yet limited in line with the smaller diffusion of information technology as a whole (Craglia, 1993): only two out of seven regions have GIS and only one has plans to acquire it in the short term.

Among the 13 Italian regions with GIS there is only one authority that has two separate and totally independent systems. In another, there are three different software integrated within a common framework. Therefore the survey identified 16 systems in total. The dominant software product is Arc/Info (8 of 16). In the other eight cases, the survey identified seven different software packages some of which are developed in Italy and distributed locally. All but one system operate on workstations, the exception being a PC based system.

Half of the systems (8 of 16) have been acquired since 1990 and are therefore at a relatively early stage of implementation. The average length of experience is 5.5 years with longer experience (10 years) among the Northern regions. In most cases, the systems are utilized by more departments in network, with authority-wide coordination organized via an ad-hoc unit under the responsibility of the Chief Planning Officer.

The first task of the regions acquiring a GIS is to produce digital maps either through aerial photogrammetry or by digitizing existing paper maps. The scale most widely chosen is 1:5000 or 1:10 000 and is thus often more detailed than that required for regional planning. However, this choice has often been influenced by the need to have maps suitable to support more detailed planning, in line with the greater links existing between regions and municipalities before the 1990 reform. The drawback of adopting these relatively large scales is that it requires a significant effort to complete the base map for the whole region to the extent that even the regions with more than 10 years of GIS experience have not yet completed this critical task. Apart from the concerns that this raises in terms of realistic updating, the lack of complete regional coverage has also inhibited the development of other applications. To speed up acquisition and updating some regions have developed, since the beginning of their GIS projects, some degree of partnership with lower tier authorities (provinces and communes). This approach is now being followed by other regions although the difficulties that exist in this type of inter-agency collaboration should not be underestimated.

Only the Piedmont region has taken a different route by developing a 1:100 000 scale map for the whole region with more detailed coverage (1:25 000) only for small portions of its territory and/or specific projects. This approach has certainly facilitated a more rapid development of regional GIS applications, particularly for environmental monitoring, but appears on the other hand to have inhibited the diffusion of GIS at the lower-tiers indicating how difficult it is to strike a balance between the needs of different levels of government.

Whilst some differences exist among the regional experiences in terms of coverage, technical specifications and standards, the most frequently developed applications include the inventory of natural resources, environmental monitoring, and some thematic mapping for forward planning. Other GIS applications are few and found only in the more established systems. They include routing, analysis of traffic flows, and the monitoring of air and water quality. In some instances, these thematic maps are not produced with the GIS but are acquired by digitizing existing paper maps and structuring their information contents in the database so that they link with other layers. They cannot be identified as 'true' GIS applications but they provide the starting point for future developments. Only in one of the 16 regional systems have the GIS functionalities been utilized more fully with thematic maps created through the buffering and overlaying functions or by extracting slope, aspect and similar data from the available digital terrain model.

In the context of this study, it is important to outline the different strategies pursued by the regions in diffusing GIS among their lower-tier authorities, provinces and municipalities. These strategies vary, reflecting regional differences in size, level of resources, and planning culture. For example, Emilia-Romagna and Veneto have chosen to pay the full cost of the GIS adopted by their provinces if they use the same software and data structure, Lombardy has encouraged its provinces to standardize their data collection and updating rather than their choice of GIS although it has financially contributed towards the provinces adopting its same software (Arc/Info). Piedmont adopted a similar strategy but appears to have failed to follow it through, while the small region of Umbria does not appear to have made any particular effort to diffuse GIS in its two provinces but has developed direct linkages with the municipal level.

Emilia-Romagna also established a number of GIS pilot projects at the municipal level consistent with the regional and provincial systems. The aim of these projects is to develop an integrated regional information system based on the detailed information collected and regularly updated by the municipalities, and fed at an increasing level of generalization to the provincial and regional authorities. This is a pioneering project developed by a region with long experience in GIS and a strong planning culture, and its implementation is being completed.

Benefits and Problems of Regional GIS

Identifying the benefits and problems of regional GIS is particularly complex as regional authorities are very large organizations and the responsibility for GIS development is often shared between internal staff acting as coordinators and customers on behalf of their departments, and outside semi-public agencies acting as consultants and technical developers (e.g. Lombardia Informatica and PIM in Lombardy, CSI in Piedmont, and CRUED in Umbria).

While little can be said about the systems that have only recently been acquired, those involved with more established regional GIS appear to be particularly concerned with organizational issues. These include difficulty of inter-departmental coordination and extremely slow decision-making within the regional authorities, difficulty in establishing information flows with other agencies, both horizontally and vertically, and changes of priorities over time. Some of these problems stem from a lack of awareness at the beginning of these projects of what was really involved, with subsequent readjustments, disillusionment, and delays. For example, most regions did not realize that choosing a relatively large scale for their base maps (1:5000–1:10000) would delay the completion of the regional coverage for well over 10 years, reduce the number of applications possible in the meantime, and make the investment difficult to justify over many budget and electoral cycles. It should be noted however, that some of these experiences go back some 15 years to the time when regional authorities had just been established, planning powers were being devolved to these new organizations, and there was little awareness of the issues involved in using IT for land-use monitoring and planning amongst both users and vendors (De Carolis, 1993).

Against this background, the direct benefits of GIS at the regional level appear to relate to the automation of map production and thematic mapping, although the lack of proper information flows amongst different levels of local government hampers the integration of operational GIS applications in the planning/management processes and limits their usefulness to decision support. There is, however, a growing recognition that the increasing responsibilities of local government in the environmental and planning fields and the pressures to improve the efficiency and transparency of the public administration require, amongst other factors, the development and utilization of information systems like GIS. Therefore, in spite of past and current difficulties the GIS diffusion process in Italian regions maintains its momentum.

GIS DIFFUSION IN THE PROVINCES

The 1990 Local Government Act has considerably extended the responsibilities of
the provinces to include financial and land-use planning, environmental management,
local infrastructure, secondary education and vocational training, public health, data
collection and processing, and technical assistance to communes in the field of
information technology. Prior to this Act some provinces, notably in Emilia-
Romagna, already had some powers in the planning and environmental fields
devolved to them by the region. This was, however, a voluntary arrangement and
by no means the norm. Therefore, the 1990 Act is of major significance for the vast
majority of provincial authorities and has stimulated many of them to acquire GIS
technology to support their new functions. In particular, the need for provinces to
prepare land-use coordinating plans for their territory (Piani Territoriali di Coordina-
mento Provinciale, or PTCP) is proving to be a major task for which GIS is perceived
as being an important supporting tool as described in this section. It must be
appreciated, however, that the 1990 Act requires some time to be fully implemented
as the provinces are still gearing up to their new tasks, and many functions have yet
to be handed over by their respective regions. Moreover, there is still significant
political instability in the country as a whole, and recruitment in the public sector
is still blocked in order to reduce the public deficit.

The background of these considerations, the diffusion of GIS in the 103
provinces was investigated through a complete telephone survey in 1993–94 which
achieved a 100% response rate. The key findings are shown in Table 6.4 indicating
that over 76% of these authorities either have GIS or are considering its acquisition.
In particular, over one third of the provinces have GIS facilities, and another 20%
have firm plans to acquire them within 1 year of the survey. Of the remaining
authorities, the survey indicates an almost even split between those having plans to
purchase a GIS in the future (21), and those having no plans to do so at this stage
(24). Given that six out of the latter authorities have only been set up following the
1990 reform of local government and have yet to be properly organized and staffed,
there seems to be little doubt of the high level of awareness of GIS among Italian
provinces.

The geographical distribution of GIS plans shows, as may be expected, that
take-up has been highest in Northern Italy, where GIS at the regional level have

Table 6.4   Plans for GIS by Italian provinces, 1993

Plans for GIS	Italy		North		Centre		South	
	No.	%	No.	%	No.	%	No.	%
Already have GIS	36	34.9	25	54.3	6	24.0	5	15.6
Have firm plans to acquire GIS within 1 year	22	21.4	5	10.9	11	44.0	6	18.8
Have plans for GIS sometime in the future	21	20.4	7	15.2	6	24.0	8	25.0
Have no plans to acquire GIS	24	23.3	9	19.6	2	8.0	13	40.6
Total	103	100.0	46	100.0	25	100.0	32	100.0

been established for some time and there is a greater availability of digital topographic data than in the rest of Italy. Whilst over half of the provinces in the North already have a GIS, and another quarter have more or less advanced acquisition plans, it is interesting to note that some 20% of the provinces indicated to have no plans to acquire GIS even though they are mostly located within regions having a GIS. An overall higher level of awareness appears in Central Italy even though only one quarter of the provinces already have a GIS. This is indicated by the high proportion of authorities having firm plans to acquire GIS within 1 year (44%) and the very low number of authorities (only 2) indicating that they have no plans for the time being. Should these acquisition plans go ahead as expected, it is possible that in the near future the diffusion of GIS in central Italy will match that in the North with over half of the provinces having this technology. In the South progress has been somewhat slower and over 40% of the provinces indicate to have no plans to acquire GIS. However, the overall level of awareness is good with 15% of the authorities having a GIS, and another 44% indicating that they have acquisition plans at varying stages of progress. Clearly the lack of any regional GIS and digital mapping coverage is inhibiting diffusion at this stage, but the survey highlighted a general commitment to introduce GIS, particularly in some regions like Sicily where two thirds of the provinces had more or less advanced plans of purchase (Figure 6.2).

The progress made by provinces in preparing their provincial land-use plans (PTCP) can be seen as a proxy to assess the extent to which the 1990 reform of local government provides an incentive to purchase a GIS. Of the Provinces with GIS 75% have their PTCP already completed or in preparation while among the 67 provinces that do not have a GIS one third (23) has completed or is preparing its PTCP. Among them, there is a very high awareness of the potential of GIS for planning applications, as 21 of the 23 are in the process of acquiring this technology. Even among the other 44 authorities which have not yet started preparing their PTCP, half declared their intention to purchase GIS as an aid to plan making and monitoring. These authorities have already devolved some planning powers from their regional councils and made it clear during the interviews that they see a strong relationship between their newly strengthened planning responsibilities and GIS.

Among the provinces, most authorities have only one GIS. Only two provinces have two separate systems located in different departments, and one province had two different software being used in the same unit. Therefore, a total of 39 systems have been purchased by the 36 authorities. Table 6.5 shows the diffusion over time of these systems and indicates that most systems (72%) have been purchased since 1990 with a reasonably steady number of 7–10 new systems per year. The average length of experience of the authorities with GIS is therefore of some 3 years. The relatively late take-up of GIS among the provinces is indicative of the very limited role that provinces had until the late 1980s and reinforces the view that the 1990 reform has had a major impact on GIS adoption at this level of government. The findings of the survey show that there isn't any significant slowing down of the current rate of diffusion. It will be interesting to monitor the progress made over the next few years to see whether the commitment expressed to acquire GIS is sustained through the current economic and political uncertainties. In terms of geographic distribution it is worth noting that during the last 2 years, the greater number of purchases has been in the Centre-South showing an increased awareness of GIS and commitment to adopt this technology.

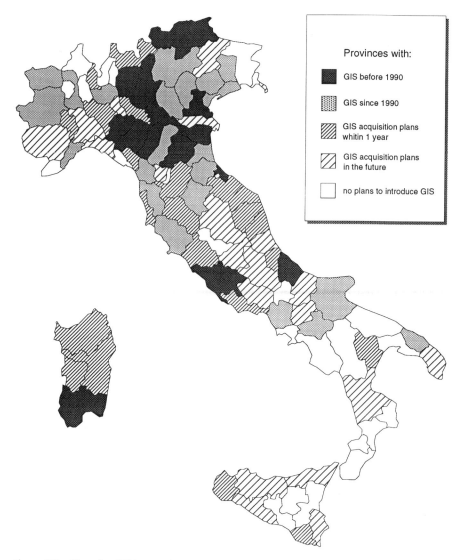

Figure 6.2 Plans for GIS by province, 1993–4.

Table 6.5 Length of experience with GIS

Year of purchase	Number	%
1993 (until July)	7	17.9
1992	10	25.7
1991	7	17.9
1990	8	20.5
1989	1	2.6
1988 or earlier	6	15.4
Total	39	100.0

Table 6.6 GIS software in Italian provinces

Software	Number	%
Arc/Info	23	59.0
Apic	3	7.7
System 9	3	7.7
Intergraph	2	5.1
Others	8	20.5
Total	39	100.0

In terms of software, by far the predominant is Arc/Info with some 60% of the market as shown in Table 6.6. However, there has been a greater diversification among the authorities that purchased their systems in 1992–93. In relation to the hardware platforms utilized, the most common configurations are workstations in UNIX environment. A significant number of authorities currently working with mainframes or PCs have also expressed the intention of moving to this configuration. It is possible that the growing experience with GIS within the public administration has started to highlight the level of human and financial resources necessary for the effective utilization of advanced systems like Arc/Info. This may explain the increasing number of choices directed towards smaller and easier to use solutions which have emerged during the last year or two.

Analysing how GIS has diffused among the provinces, three broad groups of provinces emerge depending on the type of approach adopted: 'top-down', 'bottom-up', and 'anarchist'. The first group characterized by a 'top-down' approach is by far the largest including almost half of the 36 provinces that have GIS (16). Two main forces have influenced the decisions of these authorities, regional strategies for land-use planning and GIS diffusion, and central government initiatives. Some of the regional strategies have been outlined earlier, and their impact is reflected in the finding that several of the provinces in this group belong to regions that have more established GISs. They include all nine provinces in Emilia-Romagna, two of the 11 provinces in Lombardy, and one out the seven provinces in Veneto. These figures indicate that despite the similarity in approach taken by Emilia-Romagna and Veneto local issues have played a major role in shaping the outcomes. Similarly, the financial support provided by Lombardy has not yet resulted in widespread diffusion at the provincial level, confirming that the availability of finance is a necessary but by no means sufficient condition for GIS diffusion even in areas that would otherwise appear ready for take-up.

Emilia-Romagna has clearly been the most effective in diffusing GIS among its provinces, as it is the only region in which all the provinces have GIS. This region has a strong tradition of planning in its broadest sense (land-use, economic, and social), and devolved some land-use and environmental monitoring functions to its provinces in 1984, well before the 1990 reform of local government. Emilia-Romagna was also among the first regions to adopt GIS in the early 1980s, has since completed its digital mapping coverage at 1:5000–1:10 000 scale, and purchased GIS hardware and software for all its provinces. While this commitment is remarkable and ensured a comprehensive and well-established hierarchy of plans with the technology to monitor their implementation, the GIS experience of these provinces has not been entirely trouble-free. For example, a few respondents indicated that this top-down approach inhibited their development of GIS applications.

Half of the provinces in this group share their GIS among different departments, almost always encountering organizational problems in the process. These relate to the lack of coordination and lack of agreement on the sharing of responsibilities. The lead department is almost always Planning while the IT Department appears only twice (among the 16 provinces) as a partner in the project. Other departments involved include Agriculture, Highways, and Development Control. The main applications relate to mapping, data processing, and environmental monitoring. Many of the provinces in this group have a relatively long experience with GIS and also prepared their land-use plans before the 1990 reform of local government (as in Emilia-Romagna). Therefore, they appear to identify GIS as particularly relevant to support planning implementation and environmental monitoring rather than plan preparation.

Within this group of provinces there are some interesting case-studies of the impact of central government initiatives on the diffusion of GIS in local government. In particular, a 1989 project by the now abolished Ministry for the Development of Southern Italy (Ministero per gli Interventi Straordinari nel Mezzogiorno) envisaged that local authorities in the South, among which are the 10 provinces in Campania and Puglia, should be provided with the necessary IT infrastructure for data processing and spatial analysis. The project identified GIS as the most appropriate technology and a package including a workstation, plotter, digitizer, Arc/Info software, and 1:25000 digital maps covering all 10 provinces was offered to them free of charge. Only 3 of the 10 provinces accepted this offer and one of these GIS has never been implemented. The other seven refused either for lack of awareness and skills or because they felt 'bullied into it'. This example clearly illustrates the extent to which human and organizational issues are crucial to the adoption of innovations like GIS.

The second group of provinces is characterized by a 'bottom-up' approach and includes nine authorities which decided on their own to adopt GIS and have then opted for the same configuration as their Region. They are distributed evenly across five regions without a discernible pattern. The higher degree of independence with which the decision to adopt GIS was taken may account for the higher proportion of authorities sharing the system within the authority (six of nine) than was the case in the top-down group , and also the lower proportion facing problems in doing so (three of six). Again Planning emerges as the lead department but not as clearly as in the first group (six of nine). In two cases a new department has been established with responsibility for the GIS, while in one case the lead has been taken by the IT Department. In this group there is a stronger emphasis on utilizing GIS for the preparation of land-use plans (eight of nine) while environmental monitoring is to be developed at a later stage. In fact, most of these authorities declared that GIS was purchased to support planning activities (six of nine) while 3 stated that they had originally purchased GIS for operational activities and identified planning as a useful application only at a later stage.

The 'anarchists' include 11 provinces that have decided and chosen their GIS in complete independence of other tiers of government, either because their region does not have a GIS or because technical and local considerations were viewed as more important. Of these 11 authorities five represent large cities that are also regional capitals. Therefore they have greater resources and political clout than some of the other provinces. Compared to the other two groups, there is less of a dominance of Arc/Info with some smaller packages purchased in 1992–93 either for workstation or PC. Independence of choice may account for the declared lack of problems in sharing the system among more departments which occurs in half of these provinces. There is also a greater diversity of lead departments which include the IT Department, Planning, and Development Control in this group. In terms of applications, planning and environmental monitoring are again prominent but it is worth noting that the provinces in this group have a wider range of applications and made more progress in implementing them than those in the previous two groups (top-down and bottom-up). Only among the provinces that have acquired the system within the last 12 months, is the development of applications still hampered by the lack of digital topographic data.

Benefits and Problems

Although the previous section has already indicated some of the problems identified in the survey, this section provides a wider overview of the existing trends in respect to the main benefits perceived and problems experienced in the implementation and use of GIS at the provincial level. [Compared to the questionnaire used in the British surveys (Campbell and Masser, 1991), the group of benefits labelled 'Savings' was changed to 'Increased organizational efficiency' to better reflect some of the key concerns of Italian public administration. This group was further sub-divided into: improved staff development, greater 'transparency' in the relationship with customers, cost savings, and staff savings.]

As shown in Table 6.7, just over half of the respondents identified improved information processing as the main group of benefits associated with the introduction of GIS. They were equally divided between the three types of benefits grouped under this category: greater access to data, improved data analysis and visualization, and improved data integration. Respondents who identified this as the main group of benefits, were often those with longer GIS experience. The finding that few considered GIS as improving the efficiency of their organization or providing any type of savings, is consistent with earlier work carried out in Great Britain and raises further questions

Table 6.7 Main group of benefits perceived in implementing and using GIS

Benefits	Number	%
Improved information processing facilities	19	51.4
Better quality decisions	13	35.1
Increased efficiency	4	10.8
Unable to say	1	2.7
Total[a]	37	100.0

[a] The total number of provinces considered in the analysis includes all the provinces with GIS, with the exception of Valle d'Aosta in which different tiers of government overlap, and the different opinions of the two provinces with two independent GIS.

Table 6.8 Main group of problems experienced
in implementing and using GIS

Problems	Number	%
Data related	5	13.5
Technical	3	8.1
Organizational	29	78.4
Total	37	100.0

on the claims found in the literature about the increased efficiency and effectiveness resulting from GIS adoption.

With respect to the main problems experienced by the respondents, the vast majority (78%) identified organizational issues as their main concern. In particular, lack of adequate staff and skills, resistance to innovation of staff and management, and lack of coordination among departments were the most prominent factors. Lack of adequate finance was also mentioned. Technical and data related problems were far behind as shown in Table 6.8. Among the issues cited under these headings are the incompatibility of hardware and software both within the organization and with the regional GIS, lack of data standards, and high costs of data acquisition and transfer. Only one case mentioned the problem of data security. Of all the problems cited, the lack of human resources and skills was indicated as the most crucial by over one third of all respondents. Clearly a major effort to fill this gap is required by educational institutions and vendors if the diffusion of GIS is to continue and effective utilization to be possible. Within the provincial authorities, the survey results also point to the need for a more effective organization of responsibilities to improve the effectiveness of service provision.

GIS DIFFUSION IN THE MUNICIPALITIES

Relatively few studies of GIS at the municipal level have been undertaken to date, as the large number of authorities involved (over 8000) has so far constrained comprehensive surveys at this level. To give context to the findings, it may be useful to recall that 87% of Italian municipalities have less than 10 000 inhabitants, and only some 2% have a population greater than 50 000 people, the threshold above which computing at the municipal level is most widespread (Dipartimento per la Funzione Pubblica, 1993).

At the urban level, a distinction must be made between the GIS efforts of municipalities and those of their local utility and transportation companies, as in many cases they are independent. Whilst there is evidence of considerable GIS uptake among the latter particularly in larger urban areas (see, for example, Proceedings of AM/FM Italia Conferences 1990–4), the research available suggests that the diffusion among municipalities is more limited and concentrated among middle-sized cities (50 000–250 000 population). For example Craglia (1993) identified 12 municipalities with already developed GIS in 1991 with another dozen or so having acquisition plans at various stages of implementation. The more developed systems (Table 6.9)

Table 6.9 Examples of GIS implementation at urban level, June 1991

Modena	1980	IBM-GPG Geodis
Padova	1980	Siemens-Sicad
Genoa	1983	IBM-Geomanager
Cremona	1985	Intergraph/Microstation
Perugia	1988	Siemens-ItalCad
Trento	1988	IBM-GPG Geodis II
Sesto S.Giovanni	1988	IBM-GPG Geomanager
Benevento	1988	IBM-GPG Geomanager
Lucca	1988	Siemens-Sicad
Carpi	1989	Geodis
Lumezzane	1989	Prime-System 9
Bergamo	1990	IBM Geodis II

Source: Craglia (1993).

were then investigated with a telephone survey similar to the one undertaken in Great Britain (Campbell and Masser, 1991) followed by a series of in-depth case-studies.

In terms of population size, 10 out of the 12 cities investigated are in the range 50 000–250 000 inhabitants, Lumezzane has some 24 000 and Genoa 800 000. The concentration of GIS diffusion among medium-sized cities is consistent with the finding reported earlier that computing in Italian local government is widely available only in the authorities with above 50 000 population. However, it is interesting to note that at the time of the survey none of the larger Italian cities had adopted GIS, contrary to the experience of other European countries in which there is a strong relationship between population size and extent of diffusion The only exceptions were Genoa that started developing GIS in the mid 1980s but then halted the project when key personnel left, and Turin which has developed a digital base-map in collaboration with other agencies operating in its territory as the first step towards a municipal GIS.

One explanation put forward during the interviews and the field surveys is that the fragmentation of authority among different departments and agencies in the larger Italian cities was paralysing the decision-making process in relation to large investments such as GIS. Smaller cities on the contrary require smaller investments, and can take advantage of face-to-face contacts among decision-makers to by-pass, to a certain extent, the rigidity of the administrative structure. An additional important issue among larger cities is the degree of uncertainty surrounding the establishment of twelve new Metropolitan Authorities by the 1990 reform of local government (Figure 6.3). Not all authorities have been set-up yet and their exact functions, resources, and relationships with other levels of local government (i.e. municipalities within their catchment area and provinces) are not yet clear. Despite these limitations, there are indications that the process of setting up these new authorities is increasing the awareness of the need for greater coordination and better management of information resources and exchanges among the institutions and agencies operating in these areas. Real integration of inter-agency information flows within a commonly agreed framework is however still far down the line.

These findings are supported by a recent research by Jogan and Schiavoni (1993) which focused specifically on the metropolitan areas and tried to identify all the organizations developing a GIS within these areas i.e. not only the numerous

Figure 6.3 Location of the proposed 12 new metropolitan areas.

municipal authorities in each metropolitan area (the 'metropolis' and its surrounding communes), but also provinces, utility companies and public transport agencies. Despite some gaps in the data, the overall findings of this study suggest that there are no metropolitan area-wide GIS as yet in Italy. Nevertheless, there is a significant up-take among the utilities and smaller communes. In the Centre-North of Italy for which more data are available, out of 57 organizations operating within metropolitan areas, 37 have GIS. In these areas Jogan and Schiavoni concluded that there are opportunities for the development of metropolitan GIS although some of the organizational issues highlighted above hamper such developments. In the metropolitan areas of Southern Italy investigated, these problems are particularly acute and are exacerbated by lack of human resources and common digital base-maps. Therefore, the conditions are not yet ripe for significant GIS development even though a few systems are already being developed among the utilities and other technical agencies.

Jogan and Schiavoni were not able to investigate the Metropolitan Area of Bologna which is a case of best practice in terms of the good working relationships between the different organizations operating on its territory. For example, an agreement was signed in 1994 between the Province and the Municipality of Bologna to pull together resources for the management of the metropolitan area, including the common development of GIS and other information systems. This agreement, which is open to neighbouring municipalities in the metropolitan area, shows what can be achieved through cooperation and good will. However, it must also be recognized that the strong planning tradition and the political empathy among the agencies involved in this context have favoured these developments, and are rarely found elsewhere (Secondini, 1994).

The municipal level is therefore an extremely complex environment to investigate but also one of considerable activity. The sample investigated in a previous research by Craglia (1993) remains in this context a key reference point both for its methodological consistency with the studies presented in this book, and for the window it opens on some of the key issues faced by municipalities with a longer experience of GIS implementation. The key findings are summarized below. As shown in Table 6.9, most municipalities started developing their GIS during or after 1988. Padova and Modena clearly stand out as the cities with the longest experience of GIS development, and have therefore a wider range of applications than any of the other cities. They include land-use planning, traffic and transportation, environmental monitoring, and management of open spaces.

In terms of software, there is still a significant presence of systems developed around the traditional mainframe environments (IBM and Siemens) of the municipal IT Departments which are in most cases the key partners in the development of municipal GIS together with planning departments. As the latter often lack computing experience, the choice of software has been generally made by IT Department Managers on criteria of compatibility with existing computing environments and availability of trained staff rather than functionality or ease of use. The difference between the software adopted by municipalities and that of the provinces, which are largely dominated by Arc/Info, may be explained on the one hand by the pressures on the provinces to conform to regional systems which are not felt by the municipality, and on the other hand on the less domineering position of IT Departments in the provinces as these authorities have by and large less computing experience. Therefore, provinces have been less constrained by their previous history of computing than municipalities.

The choice in the municipalities to utilize the graphical software available from the existing software supplier has often enabled the IT departments to experiment with GIS while containing the up front costs of full GIS implementation. However, the limited functionality of these software packages has often required a considerable effort by the staff involved to fill the gaps, resulting in added complexity for the non-expert user. Recently, some authorities appear to be considering a change of software (e.g. Modena from IBM to Arc/Info) to spread the use of the system through the organization and satisfy the increasing demand by technical staff who are becoming more and more familiar with computing in general and CAD systems in particular. This bottom-up demand for standardization and open, non-proprietary systems, creates additional possibilities not only for internal use but also for an increased interaction between systems adopted by different levels of government and agencies operating in the same catchment area.

In relation to the existing and planned applications, urban planning is confirmed as the main existing or future application by all the municipalities surveyed. The study also illustrates the early stage of GIS development in most municipalities. Only two cases (Padova and Modena) had already used their GIS for planning purposes; two (Lucca and Carpi) were in the process of using it to provide information and analyses for the preparation of the new city plan; one (Perugia) was at the early stages of development, while the remaining six cases were still at some distance from having a working system.

The finding that strategic planning is the main application for GIS, in spite of the lack of prior computing experience of the host departments, deserves further analysis, particularly when compared to British experiences in which operational uses like ground and road maintenance, and estate management, are much more frequent than strategic applications like plan making. However, the findings of some detailed case-studies in Italy carried out by Craglia (1993) suggest that the opportunity to initiate GIS projects was often related to the preparation of new urban plans because these normally require the acquisition of new maps. As most municipalities have to employ private mapping firms to produce the new maps, and most of these firms use digital technology, it is possible to receive the new maps directly in digital format at marginal extra cost than in analogue format. This then provides the starting point for the development of municipal GIS. In some instances, developing a GIS has also a tactical function as it enables the municipal IT departments to introduce computing in planning departments, which had previously remained unaffected by the intro-duction of computing in the local administration.

In terms of linkages, even though nine out of twelve municipalities considered belong to regions that have adopted a GIS, in only two cases were linkages found between the regional and municipal systems. In all the other instances developments tend to be independent, mainly due to organizational and political issues. Greater coordination exists between municipalities and utility companies in developing common GIS systems and sharing data (e.g. Modena, Padova, Bergamo) but it is also frequently the case where each organization has taken separate initiatives despite the fact that local utility companies are owned by the municipalities.

Among the municipalities surveyed, the main perceived benefits of developing a GIS are the possibility of having more informed technical and political decisions, followed by better information processing facilities, while savings of staff or money were never mentioned. In this respect, there is similarity between Italy and other European experiences. In terms of problems, there is general consensus that the main issues are organizational relating in particular to lack of trained staff, national legislation inhibiting the flexible management of human and financial resources and inter-departmental conflicts. Technical and data-related problems exist, but are dwarfed by the comparison with organizational ones, similarly to the findings reported at the provincial level.

CONCLUSIONS

This chapter provides a broad overview of recent GIS developments in Italian local government. Among the contextual factors which are relevant to analyse the data presented, particular emphasis has been given to the recent reform of local government (1990). This reform has redrawn the boundaries of the responsibilities of different tiers of government in respect to land-use planning, environmental monitoring, and supporting IT infrastructure. This has created a momentum for the diffusion of GIS in Italy, particularly among provinces, as well as some uncertainty at the metropolitan level. Even at this level, however, the dialogue promoted among different agencies by the process of establishing metropolitan authorities suggests that there may be a new wave of GIS diffusion in a relatively short period.

At the regional level, the findings show the central role played by the regions in developing medium-large scale digital maps. This has in some instances facilitated diffusion at the lower tiers of local government but has also significantly slowed-down the development of GIS applications within the regional authorities themselves and set the stage for considerable difficulties in updating. To address these issues of coverage and updating of regional base-maps there is an increasing trend towards inter-agency collaborations which although still limited in number point the way for future developments. The lack of GIS diffusion among the regions of Southern Italy is in contrast with the much more active role taken by the provinces in the same area and suggests that this intermediate tier of local government may play a significant role in spearheading GIS developments in the South as a whole.

At the provincial level, the research confirms that a key factor underpinning the increased diffusion of GIS is the devolution by regional councils of land-use and environmental monitoring responsibilities to the provinces resulting from the 1990 reform of local government. Whilst this legislative push appears to be more influential than the existence of regional GIS to the development of provincial systems, a factor facilitating adoption is the availability of digital topographic data which is often but not always associated with regional GIS facilities. Northern provinces have had a head-start in this respect but there have recently also been encouraging developments in the Centre-South.

At the municipal level the evidence available suggests that diffusion is clustered among the cities having 50 000–250 000 population which have the financial resources but fewer organizational problems than larger cities for introducing innovations in their organization. Planning is the prime area of application as in the other tiers of local government considered although at the municipal level the dominance of IT departments in the development of the systems has also led to the lingering on of old mainframe solutions which are more complex to use. Continued technical development and the diffusion of a CAD culture among technical staff are likely to lead to a de facto standardization of software.

Among the new metropolitan areas, which institutionally cut across the functions of municipalities and provinces, the evidence presented suggests that there are a number of GIS developments particularly among the utilities and transport agencies. However, at this stage, there is conflicting evidence as to which organization, the province or the municipality of the largest city, should take the lead in attempting to coordinate these developments. These experiences reflect the difficulties of setting up inter-agency GIS projects in complex political arenas where skilful leadership and information strategies, identifying the information flows both within and between organizations to support a given task, are needed.

The overall picture emerging from the data presented is one of rapid change both in the diffusion of GIS and in the broad context in which this takes place: political, institutional and technological. A number of important barriers have been identified among which the lack of digital topographic data is crucial. This barrier in itself underlines the lack of appropriate strategies to foster information flows among different institutional levels. Although different approaches have been highlighted: region-municipality, region-province, province-municipality, there seems to be an overall lack of concerted action based on the real information needs and resources (both financial and human) available to each participant. Other key barriers which emerge consistently in all the studies are the lack of suitably educated staff, and organizational bottlenecks both in each institution and in the broader context of Italian public administration. Critical among them are the difficulty of recruiting new staff, and of managing human and financial resources effectively. Against this background, there are however, several opportunities created by the resourcefulness and commitment of key individuals working in local government and the increasing awareness of the role that information technology, and GIS, can play in improving the quality of service in public administration. It is therefore important to continue monitoring future GIS developments in Italy.

REFERENCES

AIPA, 1994. Il Piano Stralcio 1994 messo a punto dall'AIPA, *Informatica Pubblica*, **1–2**, 6–19.

ARNAUD, A. *et al.*, 1993. The Research Agenda of the European Science Foundation's GISDATA scientific programme, *International Journal of GIS*, 7(5), 463–70.

BARRETT, S.M., 1992. Information technology and organizational culture: implementing change, paper presented at the EGPA Conference, Maastricht, August.

BEZOARI, G. and SELVINI, A., 1992. Indagine sullo Stato Attuale delle Carte Tecniche Regionali, *Documenti del Territorio*, **26**, 12–43.

CAMPBELL, H. and MASSER, I., 1991. GIS in local government: some findings from Great Britain, *International Journal of GIS*, **6**(6), 529–46

CANNAFOGLIA, C., 1993. Il Sistema Informativo dell'Amministrazione Generale del Catasto. Communication presented at AM/FM '93 Conference, Bologna, 3–4 November (not in proceedings).

CENSIS, 1990. *Informatica Italia 1989*, Milan: Franco Angeli.

CIANCARELLA, L., DURAZZI, A., MUZZARELLI, A., SECONDINI, P. and VALPREDA, E., 1992. La diffusione dei GIS nelle città europee di media e grande dimensione, in *Proceedings of 4th AM/FM Italia Conference*, Firenze, Novembre, pp..22–37.

CIANCARELLA, L., DURAZZI, A., SECONDINI, P. and VALPREDA, E., 1993. Geographical information systems in environmental analysis and planning: reasoning through a review of the principal European and Italian experiences, in *Proceedings EGIS '93*, Genova, 29 March–1 April, Vol. I, pp. 555–64.

COMA, G. and DAL FRATE, R., 1993. Il problema dell'aggiornamento del SIT catastale: l'esperienza della SOGEI, in *Proceedings of the 5th AM/FM Italia Conference*, Bologna, 3–4 November, pp. 436–43.

CRAGLIA, M., 1992. Jumping in at the deep end: GIS in Italian local government, in *Proceedings of EGIS' 92 Conference*, Munich, 23–26th March 1992, pp. 629–38.

CRAGLIA, M., 1993. Geographical information systems in Italian municipalities: a comparative analysis doctoral thesis (Sheffield: Department of Town & Regional Planning, University of Sheffield).

DE CAROLIS, G., 1993. I sistemi informativi territoriali. Luci ed ombre dell'esperienza italiana, *Urbanistica Informazioni*, **128**, marzo–aprile 1993, pp. I–XVIII.

DEPARTMENT OF THE ENVIRONMENT, 1987. *Handling Geographic Information: Report of the Committee of Enquiry chaired by Lord Chorley*, London: HMSO.

DIPARTIMENTO DELLA FUNZIONE PUBBLICA, 1993. *Rilevazione del Patrimonio Informatico della Pubblica Amministrazione*, Rome

HONEYWELL ISI., 1987. La Pubblica Amministrazione Locale, *Informatica ed Enti Locali*, **3**, 341–70.

IDC. International Data Corporation Italia, 1994. *Informatica e Pubblica Amministrazione: Prospettive e Direzioni per un Piano di Sviluppo*. Rome: IDC.

ISTAT, 1993. *13th Population Census* (1991), *Provisional Data*, Rome: ISTAT.

JOGAN, I. and SCHIAVONI, U., 1993. I sistemi informativi territoriali nelle aree metropolitane, *Urbanistica Informazioni*, **129–30**, maggio–agosto 1993; pp. I–XVI.

KING, J.L. and KRAEMER, K.L., 1985. *The Dynamics of Computing*, New York: Columbia University Press.

LIMONE, D.A., 1989. L'Informazione per il Governo Locale e la Gestione dei Servizi, *Informatica ed Enti Locali*, **1**, 11–44.

LIMONE, D., 1990. Il sistema informativo delle aree metropolitane. Risultati della ricerca promossa dall'IBM Italia, *Informatica ed Enti locali*, **VIII**(2), aprile–giugno 1990, pp. 201–5.

MASSER, I. and CAMPBELL, H., 1991. Conditions for the effective utilisation of computers in urban planning in developing countries, *Computers, Environment, and Urban Systems*, **15**, 55–67.

ORASI, A. and GARGANO, O., 1993. Territorial bases of the censuses 1991: the CENSUS Project, in *Proceedings EGIS'93*, Genova, 29 March–1 April, Vol. II, pp. 1596–7.

PAZZAGLIA, M. A., 1993. La Provincia, in *Guida Normativa 1993. Agenda per l'Amministrazione Locale*, Rome: ANCI Editioni.

RICCI, E. and SILVESTRI, M., 1993. Costi e Benefici nell'Introduzione dei Sistemi AM/FM–GIS per le Reti Elettriche di Distribuzione Pubblica: l'Analisi UNIPEDE e l'Esperienza ENEA, in *Proceedings of AM/FM '93 Conference*, Bologna, 3–4 November, pp. 10–34.

ROCCATAGLIATA, E. and PRIMI, A., 1994. A survey on GIS education in Italy, in *Proceedings of EGIS 1994*, Paris, 29 March–1 April, Vol. 1, pp. 563–71.

SECONDINI, P., 1994. Introduzione, in Provincia di Bologna (Ed.). *Progetto Territorio. La diffusione dei sistemi informativi geografici in Europa. Il caso delle aree metropolitane di media e grande dimensione*, pp. 7–20, Bologna: Centro Stampa della Provincia di Bologna.

WILLIS, J. and NUTTER, R.D., 1989. *A Survey of Skill Needs for GIS*, Liverpool: Merseyside Information Service.

CHAPTER 7

Portugal: GIS diffusion and the modernization of local government

ANTÓNIO M. ARNAUD, LIA T. VASCONCELOS
and JOÃO D. GEIRINHAS

INTRODUCTION

This chapter provides an overview of the diffusion of GIS in Portuguese local government at a crucial stage in its development. A survey was conducted between mid-1993 and mid-1994 with the support of the National Institute of Statistics (INE) supplemented by a small number of intensive interviews in the municipalities with longer experience of GIS implementation. The findings of the chapter suggest that the reported limited diffusion of GIS in Portugal and the problems experienced by end-users may be at a turning point through new legal requirements for the preparation of municipal plans and through the funds made available by recent governmental programs to support the implementation of GIS technology in Portuguese municipalities. This funding, launched and coordinated by the National Centre for Geographical Information (CNIG), promises to improve data availability, promote the use of GIS, and the benefits of the application of this technology at a local level. Favourable conditions for the implementation of local GIS have been created by CNIG which organized financial support, provided technical guidelines and assisted in selecting the systems. These programmes are strongly influencing the diffusion of GIS in the Portuguese setting.

The chapter analyses the results of the study after introducing the national context and the institutional and data environments of GIS use in the different governmental and private institutions.

NATIONAL CONTEXT

Portugal has some 10 million people distributed over 92 000 km², with two Metropolitan areas, Lisbon and Oporto, which have 2.5 and 1.5 million inhabitants respectively. Portugal is divided into 305 municipalities or *concelhos*: 275 are on the mainland grouped in five regions while 30 are in the autonomous regions of Azores and Madeira. These administrative units are further divided into districts or *freguesias*

(4208 for the whole country). Municipality size ranges from 7 to 1720 km^2 (mean area: 323 km^2). In 1991 two metropolitan authorities were established for Lisbon and Oporto with representatives from the local municipalities.

Current Environment of GIS in Local Authorities

The diffusion of GIS in Portuguese local government is taking place in a context characterized by several factors: (i) an increasing awareness of local needs and perceived advantages of data sharing, (ii) a growing number of GIS users among municipalities and the utilities, (iii) increasing data availability from private data producers, the Mapping Agencies and the Census Bureau, (iv) CNIG activities, promoting local GIS, research and training and (v) increasing functionalities of desktop mapping and PC based GIS as well as the activity of software vendors and consultants.

Local Awareness

The increasing expression of needs by local authorities reflects greater information gathered at conferences, seminars and workshops, organized by universities, research institutes and CNIG, where municipal applications often fill an important part of the programmes. The early interest in GIS had expression during the 13th Urban Data Management Symposium held in Lisbon in June 1989, which attracted more than 100 participants from the municipalities. Since then national meetings of GIS users have been held regularly (e.g. ESIG 91 and ESIG 93). The training actions organized in the last 3 years by CNIG, research institutions like UNEFOR, and the Portuguese GIS Users Association (USIG), has created a growing number of persons able to inform and influence local decision makers. Demonstrations of some of the first municipal GIS, are being shown at a growing number of meetings attended by elected and local staff and also contribute to increase the awareness of GIS in local government.

The new technological, social and political conditions developing in Portugal call for new tools to assist complex decisions involving space. There is a growing need for maps and geographical presentations as visualization tools for decision-making particularly to support the recent Municipal Master Plans (PDM). Awareness of the advantages of these technologies has been growing at the local level and the availability of financial support from central government has enhanced the demand for these systems.

The rapid changes in population and communication resulting from migration and a new expanded road network need prompt updating and fast responses. A greater demand for services in the local authorities calls for the management of huge data sets; the growth in criminality, transport problems and pollution need to be addressed by several departments in central government and within the local authorities, so calling for data integration and new tools like GIS.

Another key factor is the need to prepare new Municipal Master Plans (PDM) to comply with recent legislation. Failure to develop the municipal plan and have it approved reduces the municipal power of intervention. Therefore, a new generation of plans is emerging. Nationally there has been significant development from three approved plans in 1987 to 38 centrally approved plans at the end of 1993 and 123 at the end of 1994, the remaining being in different phases of execution. The permit issuing mechanism at the local level constitutes one of the most required areas of GIS application taking advantage of the potential for repetitive tasks with clear operational rules offered by these technologies.

Among the leading users of automated systems which are fostering the development of GIS applications at the local level are the Municipal Water Supply and Sewerage agencies (for example, in Oporto, Lisbon, Oeiras, Cascais, Matosinhos). However the trend towards the privatization of these agencies may change the way cooperation and coordination can be developed.

Automated administrative systems are already widely spread and two systems (AIRC and SIGMA) are working in a large number of municipalities with variable effectiveness. The level of use and performance of these systems needs to be analysed and its connection with studies on GIS diffusion must be fostered. These administrative systems were not designed for integration in a GIS environment and normally do not feature geocoding procedures and spatial statistics. However an evolution of these administrative systems is expected simultaneously with the growing use of GIS technology.

The new regional universities have also been an important factor in the implementation of new technologies, due to the human capital, accessibility to technology, and technical support they can provide to the municipalities.

Related Current and Potential Users of GIS

Though focusing on municipal applications, it is important to consider the main areas where GIS are in use, and the wider context of existing institutional and legal interconnections. Planning is one of the key application areas for GIS, and as mentioned earlier there are a number of important developments in this field taking place in Portugal. The current framework includes regional plans which are the responsibility of the regional commissions and provide the general directives for local planning, land use plans for protected areas under the supervision of the National Institute for Conservation of Nature (ICN) connected to the Ministry of Environment, and municipal plans (PDM) implemented at the local level (Vasconcelos and Reis, 1993). There is a general consensus as to the advantages of GIS for the development and monitoring of all these plans. Besides these plans there are two nationally defined instruments to protect sensitive lands: the National Agricultural Reserve of the Ministry of Agriculture and the National Ecological Reserve of the Ministry of Environment. Though instruments, not plans, these are strongly connected to the planning process, since both map areas with restrictions to be met at regional and local levels (Vasconcelos *et al.*, 1994).

One of the most mature systems in Portugal, within public institutions, has been operating at the Ministry of Environment—General Directorate for the Environment (DGA), since 1986. This Directorate is responsible for the National Environmental Atlas, which is being made available in several digital formats at a very reasonable cost. At 1:500 000 scale, this atlas is not detailed enough for municipal applications but is useful nevertheless to provide environmental information in standard (*de facto*) digital formats, and it is a very important initiative for the diffusion of the use of GIS functionalities in many applications and in research.

The national utilities companies for telephones, electricity and gas are already using AM/FM and starting GIS projects. These converge in the most important application areas for the municipalities. Another key area of GIS application is forestry management, since the pulp and paper companies were pioneers in GIS use. Governmental agencies, the Forestry Institute and various research institutions are also carrying out several important research projects in forestry management, fire protection and modelling.

Data Availability Issues

The National Statistical Institute (INE), the National Mapping Agency—Instituto Português de Cartografia e Cadastro (IPCC), and the Army Cartographic Services— Instituto Geográfico do Exército (IGE), together with the National Centre for Geographical Information (CNIG), the regional commissions, and the private companies are dramatically changing the availability of data in Portugal as well as increasing awareness of GIS use and benefits.

The National Statistical Institute has prepared a geographic base for spatial referencing with 106 000 census blocks, covering the whole country. That base is mostly in paper format but INE is promoting its use as a framework for data referencing. Data exchange agreements with the municipalities under a set of digitizing rules, including basic topology, aim to build up an important element of the spatial data infrastructure, a bottom-up segment based geocoding system.

The national mapping agency is making orthophotomaps at 1: 10 000 scale in conjunction with a DTM mainly for municipal plans, which now cover 65% of the country. It is also responsible for topographic mapping at 1:50000 scale and for derived cartography. The agency is changing its regulations and new legislation being prepared will regulate a simplified cadastre with a better connection between geometrical and legal cadastre, and a strong participation of the private sector. These developments will strengthen the agency's central role in coordinating data conversion and acquisition by the municipalities and utilities, and are likely to result in a much greater data availability.

Another important agency is the Army Cartographic Service which is digitizing its widely used 1:25 000 map and releasing it in vector format. Sheets from that map are also being digitized through raster to vector conversion. The DTM and hydrography are also being released. However the coverage of these products is still limited to part of the country. The Army Cartographic Service is also building a cartographic data base at medium scale using image processing for updating. It is anticipated that this will be finished in 1999.

At the local level, there is considerable pressure from daily management activities to digitize large scale base maps. This is fostering a growing number of data sharing agreements between municipalities and operators, particularly of electricity, telephones and gas, who are investing very aggressively in GIS. These agreements diminish initial data costs and facilitate integration. Municipal agencies in charge of water and sewage are also pushing in the same direction, sometimes before the municipal authority decides to adopt GIS for its other internal applications. The two metropolitan areas of Lisbon and Oporto, where one third of the population lives, seem to be where these agreements (signed or in discussion) on large scale cartography are more effective and widely spread. This cooperation is an important factor which allows for the more rapid introduction of the technology, as it results in lower costs of basic data for the municipalities.

Address data referencing systems are not in use and the quality of the address is still inadequate (see, for example, Arnaud, 1989). This presents a major barrier to natural referencing by address in most administrative files. In this context the efforts of the Post Office are of particular relevance. This agency is preparing a new postal code system, with three additional digits, mainly for distribution purposes, but it may in future improve data referencing. The Post Office is using GIS and is distributing its boundary files free of charge.

Recently completed street centre-lines have been built for the city of Lisbon, by the ESRI dealer, Octopus, and other main cities are also being digitized. Census blocks and street centre-lines have been built for the National Statistical Institute (INE) in at least three municipalities: Angra do Heroismo, Évora and Carrazeda de Ansiães and demonstrations are running on desktop mapping products.

INE is responsible for the coordination of most statistical data production and diffusion and therefore is the main source of alphanumeric data. New data collection and diffusion methodologies are being introduced to replace the traditional paper based data flows by decentralized administrative data processing under statistical data standards. INE is providing the municipalities with a computer application for tracking permits and the production of annual building statistics. This is expected to become a standard for data transfer between municipalities and INE.

The National Centre for Geographic Information (CNIG)

Aware of the importance of GIS, the Government took the lead in the second half of the 1980s and established CNIG, the National Centre for Geographical Information, to connect and coordinate the integration of data at different levels of public administration and thus develop a National System of Geographic Information. Nowadays CNIG is an entity under the Ministry of Planning and Territorial Administration. Efforts have been made in collaboration with regional agencies and municipalities to implement GIS and integrate them in the network of regional and local nodes of SNIG, the National Geographic Information Systems (CNIG, 1993). Among CNIG's activities, the two programmes outlined below have a key role in creating the background for the diffusion of GIS in local government. Both were announced in February 1994 (see Prosig and Progip, 1994).

PROGIP (Support program for the computerized management of municipal land use plans) In its creating legislation (DR33, 1994), it was recognized that effective planning increasingly requires the availability of computers to ensure access to available information in real time and as a means of inquiry, diagnosis and analysis. It follows that 'the exploitation of these tools in the management of the municipal plans of land use is a priority' which requires the existence of digital cartography and associated information relative to the current municipal plan in digital form as well as the existence of experts to work with the GIS on a full time basis, a situation not immediately feasible in all municipalities.

The aim of **PROGIP** is 'to promote the implementation in each municipality of computer tools at reduced cost and easy exploration that can allow an effective management of the plans taking advantage of the information technologies'. Moreover, the key objectives are 'to implement the computer management of the municipal land use plans and the continuous and systematic updating of information in these plans, as a way to contribute to the modernization of local administration'.

Municipalities are eligible to apply for assistance if they wish to explore these applications, provided that they assure continuous updating, accept the specialization of at least one municipal technician and support 25% of the acquisition costs of personal computers and peripherals. By late 1994, 175 municipalities had applied to PROGIP and 95 had signed agreements with CNIG.

PROSIG (Support programme for the implementation of the local nodes of the national geographical information system)   This program was established alongside PROGIP to support the development of the network for the National System of Geographical Information (SNIG). All the municipalities, or associations of municipalities, are eligible for PROSIG, provided they have digital topographic information, agree to make available on the network public domain information stored in the GIS, support the maintenance of the system, accept the specialization of at least two municipal technicians for GIS operation, support 10% of the total cost of the operation, and accept the need for exchange of experience and applications developed with the support of the PROSIG with other municipalities at no charge (DR26, 1994). By late 1994, 88 municipalities had applied to PROSIG and 46 had been selected for the first phase of the programme.

Software Vendors

The main software vendors and a growing number of consultancy companies, sometimes in connection to universities, are increasingly acting to support municipalities in data collection and conversion, feasibility studies, prototyping, applications development, systems evaluation, user requirements analysis, strategic planning, project management, training, data base design, supply of turnkey systems and maintenance. Since the launch of the PROSIG programme consulting companies have also been very active in helping the municipalities to organize their proposals.

MAIN FINDINGS FROM THE STUDY

To have an overview of GIS/AM diffusion in the municipalities prior to the launch of PROSIG and PROGIP a number of interviews were conducted in late 1993 and early 1994, in Lisboa, Barreiro, Almada, Loures and Oeiras, in the metropolitan areas of Lisbon, Matosinhos and Maia, in the metropolitan areas of Oporto, Viana do Castelo, Vila Real, Alfandega da Fé, Mirandela, Carrazeda de Ansiães, Vila Flor in the North and Évora in the South. The number of interviews does not permit statistical analysis, but the information collected was very useful for the design of a more comprehensive questionnaire and to get insight into end-users' perception of GIS and its implementation problems. The information collected through this process also complements the overall survey and provides a wealth of details not easily raised in structured types of surveys.

A brief questionnaire was mailed, in March, 1994, with the collaboration of INE, to all 305 municipalities to assess the current situation in geographical information technologies at the municipal level, focusing on geographical information systems and automated mapping. The questionnaire was developed with two requirements in mind: the need to make it short and informative while at the same time meeting the requirement for compatibility with the other national case studies presented in this volume.

The questionnaire included questions on (1) GIS: use (number, in use, testing, planning), software vendor, type of hardware used, application areas, implementation phase (conception, design, development, operation, audit), starting date, posts, comments, technical problems during system development, benefits from GIS and (2) AM: use (number, internal, external, in test, planning), type of hardware, posts, peripheral devices, existing digital maps: acquisition, type, scale, coverage.

Quantitative Results from the Survey

After some follow up phone calls (30 additional responses were received after 2 months) to the original postal questionnaire, 169 of the 305 municipalities responded (55%) and filled the part of the questionnaire on GIS. Of these municipalities 130 (76.9%) plan to acquire or are currently using GIS and 39 (23%) do not intend to implement GIS. Of the 12 municipalities with GIS, five are in the experimental phase and seven are in operation.

AM shows a higher level of adoption: seven municipalities have AM in operation, 30 are under development, 80 plan to implement, 30 do not intend to use AM and 22 gave no answer to the questions on AM. Of the 169 respondents, 131 (77.5%) are planning to implement at least one of these tools but only a small number had GIS or AM systems in operation at the time of the survey.

Due to the early stage of introduction of the technology, the municipalities were clustered in four groups: (G1) with GIS operating or under development; (G2) without GIS but with AM; (G3) planning the introduction of GIS or AM, (G4) not planning the introduction of GIS or AM. This typology reflects levels of technological absorption and expressed future intentions from the more advanced stage to the lower (Table 7.1). The first group consists of 12 municipalities (7%) and the second group has 24 (14%). Group 3 is the most numerous with 95 municipalities (56%) while the last has 38 (23%). Most of the municipalities of this last group are rural in character (35 out of the 38 classified in that group) which may suggest a lower demand for these technologies in areas where there are less development pressures.

Table 7.1 Technological absorption for urban/rural municipalities

Group	With GIS	With AM	Planning either GIS or AM	Not planning either	Non-resp.	Total municipalities
Urban (1)	6	8	17	3	20	54
Semi-urb (2)	0	5	12	0	12	29
Rural (3)	6	11	66	35	104	222
Total	12	24	95	38	136	305

1. Municipalities with one settlement > 10 000 inh. and employment in secondary sector > 38% or tertiary sector > 51%.
2. Municipalities with one settlement > 5000 inh. and employment in primary sector < 22%.
3. Others.

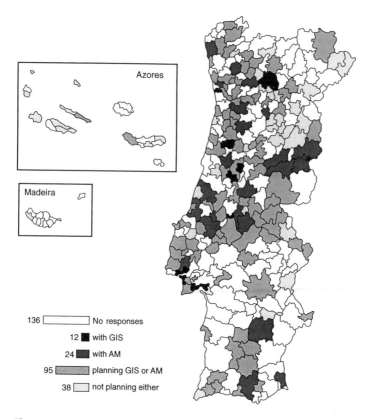

Figure 7.1   Spatial distribution of geographic information technology at municipal level.

Table 7.2 and Figure 7.1 show that geographically the first two levels seem to be more prominent in the Northern portions of the country: four out of the 12 systems in operation belong to the Northern Region, three to the Central Region and five to Lisbon and Tagus Valley region. However, the municipalities classified in group 2 have a more wide distribution: five in the North, 10 in the Centre, five in Lisbon and Tagus Valley, one in Alentejo and two in Algarve.

About 30 respondents reported that they were preparing to apply for **PROGIP** funds, and ten specified their intention to also apply for **PROSIG**. Though no question was included in the questionnaire referring to these programs, and taking into account their early stages of development, these numbers probably strongly underestimate the extent of awareness and interest in these programs, which were launched after the survey, in April and June 1994 respectively.

From Table 7.1 it is to be expected that the 95 municipalities of group 3 are potential applicants for these programs, due to their readiness to implement AM or GIS. Potential applicants also come from groups 1 and 2. From the total of 131 authorities which are potentially interested in these program not all fulfil all the conditions for application to **PROGIP** (e.g. having the PDM approved) and **PROSIG** (e.g. having digital cartography). It is hoped that these conditions will be met by a growing number of municipalities and so the diffusion of the technology will be mainly directed by these programs.

Table 7.2 Technological absorption by region

Region	Total municipalities	With GIS	%	With AM	%	Planning either GIS or AM	%	Not planning either	%	Non-respondent	%
Norte	84	4	4.8	5	6.0	29	34.6	8	9.5	38	45.2
Centro	78	3	3.9	10	12.8	22	28.2	13	16.6	30	38.5
Lisboa VT	51	5	9.8	6	11.8	21	41.2	3	5.8	16	31.4
Alentejo	46	0	0.0	1	2.2	16	34.8	3	6.5	26	56.5
Algarve	16	0	0.0	2	12.5	5	31.3	1	6.2	8	50.0
RA Azores	19	0	0.0	0	0.0	2	10.5	9	47.3	8	42.1
RA Madeira	11	0	0.0	0	0.0	0	0.0	1	9.0	10	90.0
Portugal	305	12	3.9	24	7.8	95	31.2	38	12.5	136	44.6

In this study, the responses on GIS applications reflect mostly future intentions. The number of systems reported in use is 12, while application areas are reported by 69 municipalities (53%) intending to implement AM/GIS. The application areas proposed were: automated mapping, building licensing, urban planning, demography, cadastre and property management, housing management, facilities management, municipal master plan, utilities networks, disaster prevention, land use management, built heritage and street maintenance and parking.

The order of preference given was: Master Plan preparation (55), urban planning (47), automated mapping (43), building licensing (42), utilities networks, followed by land use management and street maintenance (32). Demographic studies, cadastre, housing and urban facilities had around 22 of the 69 responses to that item, while disaster response and built heritage (15) was the less preferred. The results of this part of the questionnaire, in spite of the help provided by phone, must be used with caution as some misunderstanding of questions has been detected.

End-users' Perception of GIS Implementation

End-users' perceptions of the implementation of GIS in different types of sites is helpful to enhance awareness of hidden factors that may influence potential implementation, design strategies to cope with specific issues raised, and also identify success factors to be considered in later implementations. With this in mind, a number of intensive interviews with direct users and project managers were carried out. These brought up specific concerns, difficulties and strategies of intervention which are important for these processes. New questions emerged after these interviews. These preliminary findings already give an idea of how GIS is perceived in the Portuguese context. Some of the findings are similar to those mentioned in the literature on implementation problems. Others are specific to the Portuguese context.

For example, the municipality of Oeiras seems to be a leading authority in GIS, having the three advantages of political support, administrative compliance and funding. In fact, the municipality is ruled by a very committed President, who has been in charge for quite some time assuring political continuity. Long before its GIS implementation the municipality was well known for being very active particularly in the planning area with clearly stated objectives and careful resource management. All this was stated in the municipal master plan (PDM) which is updated regularly. Administrative compliance emerged from the possibilities opened by enacted legislation for new administrative models. Since there was already the idea of a GIS, the proposed local restructuring was aimed at being consistent with the needs of operation of this type of system. Its pioneering status in the area of GIS at the municipal level resulted in Oeiras being able to acquire considerable financial support, as the only applicant, at that time, of EU funds managed by the Regional Commission for Lisbon and Tagus Valley.

From the interviews several key issues, related to GIS implementation problems and benefits, were identified in accordance with two main components: human issues and organizational and technical issues.

Human Issues

The interviews revealed some effective and potential local GIS champions. A leader with common sense, able to develop a correct phasing for the process (user requirements, implementation plan, system and database design, system acquisition and installation, technical training, data conversion, etc.), with a precise idea of information management concepts, seems to be a key success factor. In spite of the introductory courses on GIS held by CNIG, the National GIS Users Association (USIG), and universities, along with vendor training sessions, persons aware of the main implementation steps were not always found.

More generally Autocad courses have been used in several municipalities to identify staff who are more inclined to computers and also to motivate others, thereby assessing human skills. One municipality has carried out a survey to assess the training expectations of the staff. This revealed a clear preference for computer areas. As a result a change in the assessment procedure of the department has been made, letting the staff choose further training instead of appointing new persons, to assure greater commitment.

However 'computer nuts' are seen as inadequate: 'I do not want a person who loves computers so much that it will complicate the system' said one of the coordinators of a municipal GIS expressing his disbelief that a computer expert would be appropriate for leading the implementation project.

Age was identified by several interviewees as a very important factor particularly in the eagerness to use automated systems and also for the greater flexibility to adapt to new rules. The young are much less reluctant to computer usage in general, which is of course not an unexpected finding. A key changing feature seems to be the gradual integration of a new generation which is more familiar with automated systems in the job market, particularly within municipalities. It should also be noted that staff who have already experienced the advantages of the automated systems, even for less complex tasks, were more favourably disposed to more sophisticated technology, such as GIS; so the previous existence of a computerized environment, even for administrative tasks, reduces resistance or even increases eagerness for new technologies.

Organizational and Technical Issues

The involvement of the different types of potential users in the system design is essential to remove departmental barriers but it seems a difficult task in the current organizational structures of many municipalities. However, demonstrations of the impact and benefits of new technologies may help to remove natural barriers to change.

Two cases are particularly relevant: one municipality developed an operational automated system to track the permit issuing documents within the approval mechanisms, allowing people to understand the advantages to their work of using information technology; another interviewee, with a geographical background, brought to work his personal computer to show the possibilities of using the new technology for the required day-to-day tasks.

Lack of administrative support was one of the most cited obstacles to an adequate operation of GIS. Frequently it was indicated that the hierarchical administrative structure and the concentration of power in the chief executive's office was inadequate for the development of a GIS. It was also noted that persons aware of GIS application areas and potential champions are not often in an hierarchical position with enough power to remove barriers and attain the critical results needed to convince the politicians; often they are recent employees. More flexible and decentralized administrative structures within the municipality seem to produce better results. For example, one municipality, where the organization of the GIS was already underway, took advantage of recently enacted legislation and restructured the municipal services to fit the GIS requirements.

There is a need to make people willing to dedicate initially greater amounts of time to become familiar with the system, either through easier applications or greater effort in the development of user friendly interfaces. GIS, as any other automated system, seems to have more potential for acceptance when associated with specific tasks and obvious results, particularly in the familiarization phase (PDM, permit issuing tracking). Therefore, the association of GIS with specific applications seems highly desirable to contribute to staff acceptance.

To achieve success a GIS needs effective interdepartmental and interinstitutional relations. Part of the GIS implementation problem results from the absence of habits and routines for communication and data sharing within or among institutions. GIS could play its natural integration role, enhancing these habits and contributing to local administration efficiency. Changes in traditional information flows are also being imposed by the need for effectiveness and efficiency. For example, in one municipality data from developers is collected in digital format, to allow its use in a departmental CAD system.

Overall the findings of the interviews show that internal bureaucracy, especially in the bigger municipalities, the lack of skilled personnel and delays in the acquisition of equipment are the main obstacles to the implementation of GIS. Many munici- palities also identified the geographical dispersion of different services as an obstacle to sharing data and developing a corporate system.

Many interviews expressed concerns related to the continuity of technical support from the suppliers. The high turnover rate of the employees in the companies supplying the technology was viewed as a very negative aspect, and identified as a reason to prefer specific vendors.

The need for caution was recognized due to the growing awareness in the municipalities about the cost of possible GIS failures. As a result many were hiring consulting services to carry out evaluations through pilots, before buying a system. Research institutes, in particular those linked to the recently created regional universities (Minho and Aveiro), have contributed substantially to bring the experience of new technology to distant local administrations of less developed regions.

CONCLUSIONS

The findings reported in this chapter indicate that there are strong possibilities that the national GIS profile of Portugal will change dramatically in the near future, due to the intervention of the Portuguese Government through the PROGIP and PROSIG programmes. It is hoped that GIS technology will be an extremely helpful support for the management of scarce resources in a complex and fast changing world, where decisions have to be taken on up-to-date information.

Some key factors seem to be contributing to the expected growth of GIS at the local level: (1) increasing planning activity leading to the structuring of municipal strategies for land use; (2) funds now made available, to promote investment in new technologies (e.g. PROSIG Funds, Cohesion Funds); (3) fast development, mainly due to EU funding for road infrastructure, requiring faster spatial decisions; (4) a new generation of computer educated youngster, now entering the job market; (5) the new demands of the utilities (electricity, telephones, water) forcing data providers to adopt new technologies and improve data availability through data sharing agreements; (6) the creation of the National Centre for Geographical Information showing central government commitment to coordinate efforts.

The findings of this empirical study, which was carried out before the launch of PROSIG and PROGIP, may be a reference point against which to compare the diffusion process of GIS, after the first round of local GIS supported by government funds is implemented. Regular follow-up surveys will be needed in future to monitor the effectiveness and efficiency of the changes induced by the introduction of the technology and evaluate the extent of implementation. In this way important information will be collected on GIS diffusion at the local level which will provide further information to the central coordination agency, helping it identify and address barriers and to promote the effective use of the technology.

For some municipalities, land-use planning may be the first level of GIS application, with information on census blocks and road networks which allows traffic analysis, ecological and agriculture zoning, and socio-economic analysis. This level may be reached with easy to use desktop mapping software, with low investment in data, equipment and training. This first level could also quickly provide a working demonstration of the benefits of the technology, and a means for future user needs assessment prior to the implementation of a full GIS covering more demanding application areas, or even a fully integrated GIS.

Several implementation strategies may be proposed for evaluation considering that there are several classes of uses of GIS/AM in a local authority: traditional maps for planning and decision making (desktop mapping), utility networks and public works management and operations (AM/FM), computer aided design and drafting packages for engineers and architects (CAD), while other users require more sophisticated systems such as property management linking databases with mapping facilities or GIS linking data from different sources through geographic location.

Municipalities considering the use of GIS technology may then decide to adopt different implementation strategies, through departmental applications or an integrated GIS, according to their type, size, location, organizational structure, availability of data and resources. The advantages and disadvantages of corporate and departmental approaches have been well explained in the UK (Campbell, 1993).

In any event the ability of GIS to link separate databases may cause changes in the activities performed by different departments and changes in the organizational structure may be needed, leading to the re-engineering of local government. GIS have the potential for greater data sharing among departments, improving data consistency and better communications between departments and other governmental agencies.

GIS can integrate all locational information, collected and managed by the local authority as well as from external sources but address standards and data quality controls are needed to guarantee the usefulness of the system. The need for standards and rules for all institutions producing digital data is essential to facilitate the use and sharing of information. The regulations produced by the national mapping agency (IPCC) will be updated soon, in order to include rules and data structures for use in GIS environment. The National Statistical Institute (INE) has developed a set of rules for municipalities who wish to digitize the census boundaries and CNIG has produced a set of guidelines to implement municipal GIS. It is expected that current efforts on standardization at the European level by a variety of organizations such as the European association of national mapping agencies (CERCO) and Technical Committee 287 will contribute to an easier integration of geographic information at the European, national and local level. A set of new regulations is expected, minimizing waste in data collecting and conversion, proposing data sharing among and within organizations, fostering information flow and its effectiveness (for example, balanced costs, transfer formats, time effects) and promoting the build up of a Spatial Information Infrastructure. At the national level it is now also possible for countries introducing GIS to learn from the experiences of earlier adopters. This may lead to more rapid access to the benefits of geographic information technologies and take advantage from the increased functionality of the available software and new data collecting methods whilst decreasing implementation costs.

REFERENCES

ARNAUD, A.M., 1989. A step-by-step geocoding strategy for local planning, OECD Seminar, *Coordinated Information Systems for Improving Urban Functioning and Management*, Copenhagen, Denmark.

CAMPBELL, H., 1993. GIS implementation in British local government, in Masser, I. and Onsrud, H. (Eds) *Diffusion and Use of Geographic Information Technologies*, Dordrecht: Kluwer.

CAMPBELL, H. and MASSER, I., 1992. GIS in local government: some findings from Great Britain, *International Journal of Geographical Information Systems*, 6, 529–46.

CNIG, 1993, *Inventário de SIG Comercializados em Portugal*, Boletim 1.

DR26: DESPACHO MINISTERIAL, II Series, 1 February, 1994, PROSIG, 'Programa de Apoio à Criação de Nós Locais do Sistema Nacional de Informação Geográfica'.

DR33: DESPACHO MINISTERIAL, II Series, 9 February 1994, PROSIG, 'Programa de Apoio à Gestão Informatizada dos Planos Municipais de Ordenamento do Território'.

DEPARTMENT OF THE ENVIRONMENT, 1987. *Handling of Geographic Information: Report of the Committee of Enquiry Chaired by Lord Chorley*, London: HMSO.

MASSER, I. and CAMPBELL, H., 1995. Information Sharing: the effect of GIS on British local government, in Onsrud, H. and Rushton, G. (Eds) *Sharing Geographic Information*, New Brunswick, New Jersey: Rutgers.

'PROSIG E PROGIP', 1994. Editorial in *Boletim de Informacao Geografica*, 5, 4–5, Lisbon: USIG.

VASCONCELOS, L.T. and REIS, A.C., 1993. The Portuguese Planning Process —detailed at the local level, paper presented at University of Newcastle upon Tyne, 15-17 October.

VASCONCELOS, L.T., GEIRINHAS, J. and ARNAUD, A.M., 1994. Tecnologias de Informação Geográfica nos Municípios, *Boletim de Informacao Geografica*, 5, 4–5, Lisbon: USIG.

CHAPTER 8

Denmark: Local autonomy, register based information systems and GIS

HANS KIIB

INTRODUCTION

This chapter presents the findings of a comprehensive telephone survey conducted at the beginning of 1993 in the context of the Danish research programme on the Diffusion and Impact of GIS on Local Government in Denmark. The survey was carried out in a comparative framework with the studies in Great Britain and covered all 275 municipalities and 14 counties in Denmark. The findings provide an overview to the type and size of local authorities, geographical location, take-up over time, software and hardware utilized, use of the technology, and the perceived problems and benefits associated with GIS. Particular attention is given to the change in implementation strategies related to the different kinds of geographic information technologies during the last two decades, from centralized register-based systems to distributed AM systems and GIS, with a view to developing a typology of implementation strategies for GIS in Danish local government. To give context to the research findings, the following sections introduce the structure of Danish local government and its use of information technology.

THE INSTITUTIONAL CONTEXT

Denmark is a small country with 5 million inhabitants and three administrative levels of government: Central government, 14 counties and 275 municipalities (Figure 8.1). The history of modern local government in Denmark goes back to the development of the post-war Welfare State. At the end of the 1960s a comprehensive municipal reform was implemented in Denmark as in many other northern European countries (Great Britain, The Netherlands, Belgium, Sweden, Norway). About 1300 municipalities were aggregated to 275 larger municipalities. The number of inhabitants of Danish municipalities averages 18 500. All counties have more than 200 000 inhabitants with the exception of the county Bornholm (island in the Baltic) with only 47 000 inhabitants. The average size of the counties is 350 000 inhabitants.

Local authorities are generally under the technical supervision of the Ministry of

Figure 8.1   Danish local government including 14 counties and 275 municipalities.

Interior but since the local government reform of 1970 municipalities and the new county authorities have gradually increased in importance forming a system of local self-government to which considerable power is devolved. The successive transfer of functions away from central government has concentrated the major part of the local and the regional functions in local government. New public functions are almost invariably performed by municipalities and, to a lesser extent, counties. This has led to the development of local government as an all-purpose unitary administration, where broadly speaking all local functions are administered or supervised by a municipal council. In no other Danish authority are such wide administrative powers vested as in municipal councils. In principle they deal with all domestic issues for which the public sector is responsible, including tax assessment, help of people in distress, urban planning, land management, construction of local roads and water supply and air purification, emergency service and civil defence.

A comparative analysis of local government identified Denmark as having one of the most decentralized public administrations in Europe (Batley and Stoker, 1991). Compared to other countries Danish local government plays a major role in government and administration of functions and issues for which the public sector is responsible. Danish local government is rather small but has a strong economic power as 70–80% of local government income is based on municipal taxes (personal tax and property tax) and the level of local government expenditure as a percentage of GNP is three times as high as in Great Britain (Mouritzen and Nielsen, 1988). The level of local government employment as a percentage of total national labour force is 16.8 in Denmark and only 8.5 in Great Britain (Batley and Stoker, 1991;

Nielsen, 1991); reflecting the increased number of functions transferred to local government in Denmark during the last 25 years (Mathiesen, 1983). To carry out these functions, local government in Denmark has also developed a high degree of inter-municipal cooperation in the supply of services and infrastructure.

Local authorities are the main providers of large scale maps. They control, plan and manage the utilities, and are co-managers of the Land Registry and the Cadastral Register. They also produce and manage (coordinated by central government) coherent and comprehensive register information related to land parcels, property, land zoning, taxation, person related information and company related information. In contrast, the functions of the counties are concentrated on hospitals and public health care; rural planning, energy planning and environmental planning and monitoring; highways; and secondary schools (Gymnasium/A levels).

LOCAL GOVERNMENT USE OF INFORMATION TECHNOLOGY

In the Danish post-war context the major reforms related to functions of local government or reforms related to major changes in economic conditions have often been followed by change in the administrative structure, by change in the internal organization of work in local government and by *implementation of new technology*. This has certainly been the impact of the major municipality reforms in the early 1970s and to some extent also later in the 1980s where major energy reforms and environmental reforms pushed new responsibilities to the counties and municipalities. Thus this section discusses two waves of information technology related to the diffusion of GIS in local government in the 1990s:

- The first wave (1970–85) is *Registers and Register Based Information Systems*;
- The second wave (1985–91) is *Digital Mapping and Facility Management*.

Registers and Register Based Information Systems

The first wave of geographic information technology in Danish local government dates back to the early 1970s when a tax reform, a social policy reform and a planning reform were introduced. The municipalities then became responsible for the major part of the public collection of taxes, the transfer of income, urban planning and land use management. Integrated computer systems and routines were established including a range of digital registers and administrative information systems. In 1972 'Kommune Data' was founded as an inter-municipal enterprise, based on non-profit partnership. Kommune Data and 'Datacentralen', which was the data manager of Central Government, were made responsible for the establishment and management of the registers, based on mainframe and on-line terminal solutions. But the municipalities and the counties were made responsible for data collection and update. Kommune Data developed a range of information systems based on mainframe solutions and most municipalities were linked to these and made use of them particularly in the areas of taxation, social services and health services.

It became the rule of thumb that almost every new public function should be followed by a new administrative information system and to a large extent new registers were developed as well. In the 1970s a range of digital registers were established: The Population Register; The Property and Valuation Register; The Building and Dwelling Register and The Enterprise Register. During the 1980s The Planning Register and The Natural Resources Data Register were established and a number of manual registers in central government were computerized including The Cadastral Register. Due to the very high degree of integration of the efforts carried out at central and local government levels there is often a corresponding integrity of the data in the registers. Therefore, it is possible to take these 'old' registers as a basic reference, common to all local authorities and administrations in local government in Denmark today. Central digital registers and register-based systems have become a basic administrative tool in departments related to economy, income and property taxation, and social care, while departments in charge of environmental planning, land management and technical affairs tend to use more frequently their own departmental registers (Boligministeriet, 1986). And it is here in the technical departments that the need for new mapping and management systems has been formulated most strongly.

Digital Mapping and Facility Management

From the late 1970s to the mid-1980s a second wave of reforms concerning energy supply, energy conservation and environmental audit were passed by parliament adding another big sector to the activities of counties and municipalities. Especially big investments in energy supply and the establishment of the regional natural gas companies led to a new perception of Land Information. These companies wanted to build up their information systems as a Land Information System (LIS), linking the large scale digital topographic map and the cadastre to the register information. They wanted to build up LIS for handling: technical and user-related information— such as data to plan energy supplies and utility mapping to document the position and performance of the network—and economic and user-related information— including for example data for creating budgets, consumer excise duty accounting and calculation of consumer prices.

This situation forced the two main producers and suppliers of large scale maps, the municipalities—in co-operation with photogrammetric firms—as well as the National Cadastral Administration to change from manual cartography to digital mapping. During the 1980s a number of pilot projects took place and the first Danish digital map format, the DSFL format, was established in 1987 (Brande-Lavridsen, 1989). In the same year the Danish Survey and Cadastre was established (Larsen, 1989) and a strong political awareness of the need for a coherent national information management policy started to emerge. The objectives of this new policy included:

- converting a number of nation-wide basic registers concerning the built environ-ment into a digital format (for example the Cadastre and the Land Register);

- integrating the digital topographical map as well as the cadastral maps with the register based information systems; and

- opening an integrated information system at large, the Co-ordinated Information System (CIS) to citizens, firms and public in general (Trollegård, 1992).

SURVEY RESULTS

The description above provides an overall framework for the local government adoption of pre-GIS technology. The survey of 1993 on the diffusion of GIS in local government in Denmark provides information on:

- the use of register based information systems in local government today;
- the historical and geographical patterns of the adoption of AM/FM as well as GIS in local government;
- the present adoption and use of GIS;
- the software and hardware being utilized;
- the perceived problems and benefits associated with GIS;
- the future plans related to the implementation of GIS in local government.

The Present Use of Register Based Information Systems in Local Government

Today all municipalities have access to a range of digital registers. The registers themselves serve as information and management systems in the 'home'-departments (tax, social services, building permission) of the local government who are also responsible for the updating of these systems and providing up-to-date geographical information on many issues related to planning and management. Besides, Kommune Data and private firms offer a wide range of applications without maps, including management systems and decision support systems on top of the register information.

The survey indicates (Table 8.1 and Figure 8.2) that over 80% of all municipalities are using one or more register based information systems where the georeference is present (Address, Parcel Number/Cadastral ID, Building and Dwelling ID). All municipalities over 25 000 inhabitants and 74% of the smallest group under 10 000 inhabitants are using that kind of application, run on mainframes, workstations, PCs or terminals to Kommune Data. These figures show the level of adoption of pre-GIS information technology in local government.

Table 8.1 Information systems in Danish local government, January 1993

	Local authority		GIS		AM/FM		Register based inf. system	
	No.	%	No.	%	No.	%	No.	%
Municipality								
>100 000	5	2	2	40	3	60	5	100
50–100 000	9	3	6	67	3	33	9	100
25–50 000	31	11	10	32	15	48	31	100
10–25 000	92	33	13	14	35	38	73	79
<10 000	138	50	2	1	31	22	102	74
Total municipalities	275	100	33	12	87	32	220	80
Counties	14	100	6	43	3	21	3	21

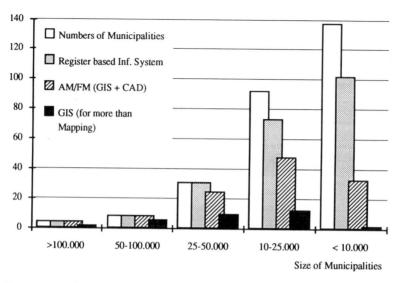

Figure 8.2 Adoption of information systems in Danish municipalities, January 1993, related to the size of local government.

Digital Mapping and GIS

The level of adoption of GIS and GIS related AM/FM technologies in local government in Denmark (Jan. 1993) can be seen from Table 8.2. The second column shows that 33 (12%) municipalities and 6 (43%) counties had introduced GIS technology and used it for more than just mapping purposes. In all, 87 municipalities did not consider their system, primarily used as an AM/FM, as 'real GIS'. In these municipalities there are 90 map handling systems of which at least 34 are based on GIS technology (the third column in Table 8.2). In the fourth column these figures have been linked to the figures of the second column . This shows that the present adoption of GIS in local government is 24% in municipalities and 43% in counties; 71% of the municipalities with over 25 000 inhabitants, 25% of the municipalities with between 10 000 and 25 000 inhabitants and only 8% with under 10 000 inhabitants had implemented GIS before 1993.

As shown in Table 8.2, column 5, the total level of AM/FM and GIS systems is high in Danish municipalities and counties; 44% of all municipalities and 64% of the counties had adopted a GIS or an AM/FM before 1993. These findings show a clear link to the general context in which the diffusion process is embedded: the municipalities are the main providers of large scale maps. The counties do not have this role. Here the use of GIS for basic mapping is more limited.

Table 8.2 displays a close relation between the size of the local authority and the adoption of the GIS technology showing that the larger municipalities have by far the highest level of GIS adoption. Figure 8.3 gives a visual impression of this. It also shows that a great proportion of the municipalities under 25 000 inhabitants for various reasons have preferred to invest in CAD based systems for handling the digital maps and map related information.

Municipalities over 25 000 inhabitants have a strong economic capacity and thus the ability of local investments in advanced information technology. They are

Table 8.2   AM/FM and GIS in local government in Denmark, January 1993

	Local authority		GIS[a]		AM/FM[b] GIS based		GIS[c] in local govern.		GIS[c] + AM/FM[d]	
	No.	%	No.	%	No.	%	No.	%	No.	%
Municipality										
>100 000	5	2	2	40	3	60	5	100	5	100
50–100 000	9	3	6	67	2	22	8	89	9	100
25–50 000	31	11	10	32	9	29	19	61	25	81
10–25 000	92	33	13	14	10	11	23	25	48	52
<10 000	138	50	2	1	10	7	12	9	33	24
Total municipalities	275	100	33	12	34	12	67	24	120	44
Counties	14	100	6	43	0	0	6	43	9	64

[a] GIS used for more than mapping.
[b] AM/FM based on GIS technology.
[c] GIS technology.
[d] AM/FM based on CAD.

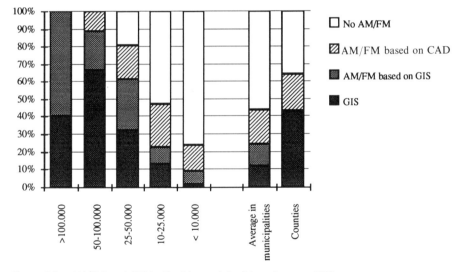

Figure 8.3   AM/FM and GIS in Danish municipalities , January 1993.

responsible for the planning and maintenance of technical infrastructure including supply systems for water, power and central heating. These functions require a close technical as well as geographical co-ordination and the advantages of doing this within a GIS may have convinced larger municipalities to invest in this technology.

Small municipalities have the same responsibilities but they often contract local surveyors, photogrammetric firms or local mapping co-operatives (owned by a group of municipalities) for large scale mapping. Water or heating supplies are often a concern of the individual or of the community, run by local water associations, local heating associations, the regional gas companies. This indicates that smaller munici-palities with a lower economic basis and less skilled capacity prefer to invest in cheaper technology like CAD based systems for the handling of maps and that to some extent they prefer to take a joint approach to large scale mapping together with the association of the water and heating supply companies.

Rate of Adoption of AM/FM and GIS in Local Government and Future Plans

The awareness of the possibilities of GIS is high in Danish local government. Figure 8.4 shows a big increase in the adoption of GIS as well as AM/FM in the period from 1987 to 1992. The level of GIS adoption rose from about 3% per year in 1987–88 to about 13% in 1991 and 8% in 1992. The figures from the AM/FM adoption are even higher; from about 6% per year in 1987–88 to 15% in 1991 and 32% in 1992. The middle size municipalities are the main investors in the GIS technology in 1991–92 and the small size municipalities (under 10 000 inhabitants) which represent over 50% of local government, are the main group in the adoption of GIS or CAD based AM/FM applications.

Some of the comments from the respondents during the interviews indicate that many smaller municipalities have decided 'to start with a map-handling system', in order to gain some experience with mapping and the use of the digital maps in the management of pipes and in land use planning. Most respondents are aware of the possibility of expanding AM/FM into a 'real land information system (LIS)', linking the register information to the various objects in the digital maps. Many vendors of CAD or GIS based AM/FM provide extension units which allow map-users access to databases containing attribute data. So many small municipalities have seen the cheaper CAD systems as a vehicle for the achievement of a LIS or GIS. The figures related to the counties are similar to those for the large municipalities. They prefer to invest in GIS either because they are going to convert manual registers or because they are going to build up new databases in areas such as health care, rural planning, environmental monitoring and planning.

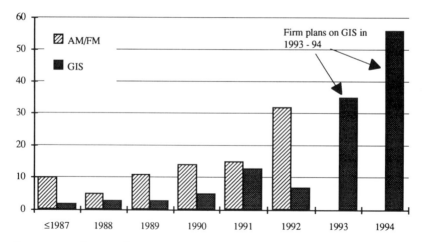

Figure 8.4 Adoption of GIS and future plans in local government in Denmark, January 1993.

Table 8.3 GIS and plans for GIS adoption in local government in Denmark, January 1993

	GIS[a]		GIS[a] 1993–94		Considering GIS[a]		No plans		Predicted GIS[a] in 1994	
	No.	%	No.	%	No.	%	No.	%	No.	%
Municipality										
>100 000	2	40	1	20	1	20	1	20	3	60
50–100 000	6	67	3	33	0	0	0	0	9	100
25–50 000	10	32	13	42	7	23	1	3	23	74
10–25 000	13	14	33	36	39	42	7	8	46	50
<10 000	2	1	46	33	65	47	25	18	48	35
Total municipalities	33	12	96	35	112	41	34	12	129	47
Counties	6	43	3	21	5	36	0	0	9	64

[a] GIS used for more than mapping.

These comments are to some extent confirmed by the figures in Table 8.3 about future plans. About one third or 35% of all municipalities have firm plans to adopt a GIS in 1993–94. Looking at the different sizes of local government this is almost the same level in all groups with the highest level (42%) in the middle sized group and the lowest level (33%) in the group of small municipalities. A rather higher proportion (40–60%) of local government with AM/FM have firm plans for GIS. If these plans for GIS adoption are carried out as predicted in the survey the level of GIS implemented in the municipalities in Denmark will be about 47% at the end of 1994, of which municipalities with under 10 000 inhabitants will have a level of 35%. Of the other half (146 or 53%) made up by mainly small municipalities, 41% is considering the acquisition of GIS and only 34 (12%) do not have any plans to introduce GIS or have not answered the question. The figures of expectations in the counties are even higher as the predicted level of GIS adoption before 1995 is 64%.

Research from Great Britain in 1993 (Campbell, 1993) indicates that the prediction of the 1991 survey on local government adoption of GIS was not entirely realized due to uncertainty about the future structure of local government and the decrease in the financial resources of British local government. It is possible that the adoption of GIS in Danish local government will be influenced to some extent by similar factors or by competitive development of more user-friendly register based information systems for local users . But even with an adoption of GIS at a level of 50% of the predicted level there will still be a tremendous increase in GIS implementation in Danish local government. In terms of the figures it comes close to 500% over the next 2 years (or about 200% each year).

Software and Hardware

The survey findings show a wide range of software products, from very advanced GIS packages to relatively simpler mapping ones. However, two American and two

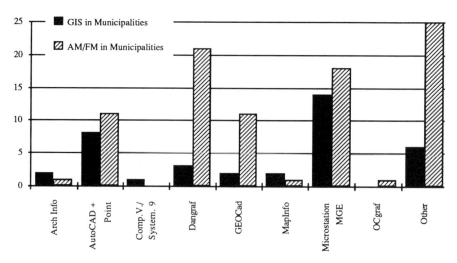

Figure 8.5 GIS and AM/FM software in municipalities, January 1993.

Danish products account for nearly 80% of the market in the municipalities (Figure 8.5). Among the Danish products Dangraf (developed by Kommundata and Jydsk Telephone) and GEOCad have a relatively small share of the municipal GIS market (9% and 6% respectively) but are strong among AM/FM installations (27% and 14% respectively). In this area of application they have a very similar market share to the two North-American products Intergraph Microstation (MGE) and Autocad-LIGS (23% and 14% respectively) which however are more significant for their dominance of the GIS market reaching 42% and 24% respectively of municipal installation.

It is notable that Arc/Info has a very limited share of the market compared with other Scandinavian countries (Andersson, 1991), Italy and Greece where this software is dominant. The situation in Denmark probably has a very local explanation due to the lack of a Danish Arc/Info vendor for many years. However, it is likely that the market share will change. One reason is, that Kommune Data has decided to stop the development of Dangraf and instead develop their GIS applications based on Arc/Info. It is also likely that PC based Desk Top Publishing and GIS packages will meet the demand of planning departments in smaller municipalities.

In the counties, the adoption of GIS and AM/FM is almost entirely concentrated on Integraph Microstation (MGE) and MapInfo packages. As far as hardware is concerned (Table 8.4) 49% of the GIS systems in the municipalities run on PCs and 44% on workstations. The breakdown of the figures of the hardware acquired by different sizes of municipalities suggests that municipalities with over 25 000 inhabitants have implemented workstation solutions and smaller municipalities have preferred to start on PC platforms.

GIS Adoption and Plans for GIS by Region

Table 8.5 analyses the findings with respect to the regional distribution of GIS facilities (third column), AM/FM (fourth column) and register based information systems (fifth column). The figures in this table are related to the diffusion of

Table 8.4 Use of hardware by type and size of local government

	Has GIS[a]		Mainframe		Workstations		PCs	
	No.	%	No.	%	No.	%	No.	%
Municipality								
>100 000	2	3	1	33	2	67	0	0
50–100 000	6	7	1	14	4	57	2	29
25–50 000	10	12	0	0	8	67	4	33
10–25 000	13	15	1	7	3	20	11	73
<10 000	2	2	0	0	0	0	2	100
Total municipalities	33	39	3	8	17	44	19	49
Counties	6	6	1	17	2	33	3	50

[a] GIS used for more than mapping.

Table 8.5 Information systems in local government by county

County	Municipalities No.	GIS		AM/FM		Register based information systems	
		No.	%	No.	%	No.	%
Nordjylland	27	3	11	9	33	19	70
Viborg	17	3	18	3	18	16	94
Ribe	14	2	14	4	29	14	100
Sønderjylland	23	1	4	11	48	16	70
Vestsjælland	23	1	4	4	17	16	70
Storstrøm	24	1	4	4	17	15	63
Bornholm	5	0	0	1	20	5	100
Aarhus	26	1	4	11	42	18	69
Vejle	16	5	31	4	25	15	94
Ringkøbing	18	4	22	5	28	17	94
Fyn	32	2	6	14	44	25	78
Roskilde	11	1	9	5	45	10	91
Frederiksborg	19	3	16	4	21	16	84
København	20	6	30	8	40	17	85
Total	275	33	12	87	32	219	80

information technology in municipalities by county. It is very difficult to talk about rich and poor regions in Denmark today or to talk about well developed and less developed regions in terms of industrialization, development of service or rate of unemployment. But it is still possible to distinguish between 'urbanized' regions with large municipalities and less 'urbanized' regions with many small municipalities and large rural areas. Traditionally the region of Copenhagen (København, Roskilde and

Fredensborg), the eastern regions of Jutland (Aarhus and Vejle) and the island of Fyn has been recognized as the most wealthy and most strongly industrialized. To this group of regions the western part of Jutland (Ringkøbing) is added, because this region has had a major development of industry and service industry in the last two decades. The largest cities and the highest level of personal income are located in this group of counties. The less 'urbanized' regions are located in the Northern and the Southern parts of Jutland, in the Western and the Southern parts of Sealand and the island of Bornholm.

The findings from the survey provide one clear regional pattern in the GIS acquisition: the 'urbanized' regions with large cites have a higher rate of adoption of GIS and AM/FM than the rest. From Table 8.5 it can be seen that three counties (Copenhagen, Vejle and Ringkøbing) include 15 (or 45%) of the 33 GIS municipalities. The first two counties represent old 'urban regions' and the third county includes some of the municipalities in the new industrial region. By dividing the 14 counties into two groups with the seven most 'urbanized' counties in one group, the survey indicates that this group represents two-thirds of the municipalities with a GIS and 60% of the municipalities with an AM/FM mapping system. Two regions have a very high proportion of municipalities with mapping systems. These are the island of Fyn and Sønderjylland (South Jutland), where mapping systems are available in half of the municipalities. In these counties there are also two of the oldest inter-municipal digital mapping corporations (Kortcenter Fyn and IGS Center), which have had a strong impact on the diffusion of mapping systems in these regions. A new inter-municipal centre has been established in Vestsjælland and these centres will play a central role in the future adoption of GIS in the three regions with respect to common standards and choice of software. Table 8.6 indicates that the same group

Table 8.6   Future plans on GIS adoption by regions

County	Municipalities with GIS, 1993		Firm plans on GIS adoption 1993–94		Considering GIS		No plans		Predicted GIS in 1994	
	No.	%	No.	%	No.	%	No.	%	No.	%
Nordjylland	3	11	9	33	16	59	0	0	12	44
Viborg	3	18	3	18	13	76	0	0	6	35
Ribe	2	14	4	29	12	86	0	0	6	43
Sønderjylland	1	4	11	48	15	65	0	0	12	52
Vestsjælland	1	4	4	17	15	65	0	0	5	22
Storstrøm	1	4	4	17	14	58	0	0	5	21
Bornholm	0	0	1	20	5	100	0	0	1	20
Aarhus	1	4	11	42	17	65	1	4	12	46
Vejle	5	31	4	25	10	63	0	0	9	56
Ringkøbing	4	22	5	28	13	72	0	0	9	50
Fyn	2	6	14	44	22	69	0	0	16	50
Roskilde	1	9	5	45	9	82	1	9	6	55
Frederiksborg	3	16	4	21	13	68	2	11	7	37
København	6	30	8	40	11	55	1	5	14	70
Total	33	12	87	32	185	67	5	2	120	44

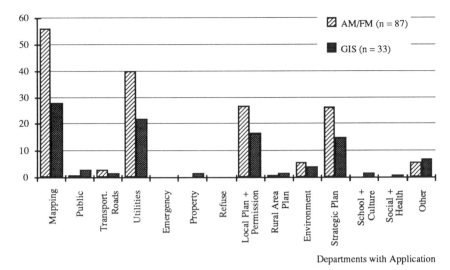

Departments with Application

Figure 8.6   Use of AM/FM and GIS in municipalities in Denmark, January 1993.

of (seven) 'urbanized' counties are the leaders in terms of acquisition plans of GIS
for 1993–94 relative to the total number of authorities.

Use of GIS

In all, 33 municipalities and six counties have implemented GIS and a further 87
municipalities and three counties have implemented an AM/FM system based on
GIS or CAD technology. The respondents in these municipalities have been asked
about the departments or units which are using or have access to the system.

As might be expected most systems in the municipalities are used for mapping
(including production, editing and hard copy printing). Figure 8.6 shows that 28
(85%) are using their GIS and 57 (72%) their AM/FM in the *mapping unit*. Apart
from the use for mapping or the handling of the digital maps there are notably only
four areas where there is a significant use of AM/FM and GIS in the municipalities.
These are:

- *utilities*; this unit includes staff related to the water supply, the heating supply, the
  management of sewerage and in some big municipalities also the electricity supply.
  Two out of three (67%) of the GIS municipalities are using GIS in these units and
  in 50% of the municipalities with AM/FM the mapping system is used in the units
  of utilities;

- *land management*; this unit includes the staff related to planning permission
  (Danish local plan system), building permission and local management of the
  cadastre. Here the figures are 50% (GIS) and 35% (AM/FM);

- *strategic planning*; this unit includes staff related to comprehensive municipal
  planning, strategic land use planning, coordination between the municipal plan
  and the technical sector plans, the social sector plans and sector plans of schools
  and culture. Here the figures are 45% (GIS) and 33% (AM/FM);

- *environmental planning*; 12% of the GIS installations and 8% of AM/FM are used in environmental planning.

All these units belong to the Technical Department. Only the unit for strategic planning has a different organizational position in some larger municipalities where it may be a part of the Chief Executive's Department or the Department of Economic Planning.

The survey does not suggest whether the systems are functioning well or to what extent and for which specific operations they are used. But out of the comments from the respondents it appears that the systems are mostly used for handling digital maps and providing hard copies, technical/administrative management of pipes, cables and other technical infrastructure, and providing a graphical representation of land use planning, urban renewal, planning and building permission in digital format. It also appears that GIS as well as AM/FM systems are used as a tool for easy access to detailed information about parcels, buildings or technical installations and that the technology is not used as a data modelling tool in local government.

In the counties the pattern is very much the same. The main users are located in the units of mapping, utilities and the units of environmental planning and rural area planning. Comments from the respondents state that their primary use is in the handling of digital maps for hard copy printing and for rural and environmental planning. The GIS in the management of roads and utilities is very much at an early stage of implementation.

Benefits and Problems Associated with the Implementation of GIS

This section examines the wider implications of system implementation in terms of perceived benefits and main problems related to the implementation process and the use of the systems. Respondents were asked in both instances to rank three sets of factors in order of importance and then to describe in more detail the group of benefits and problems which they ranked first. The results of this part of the survey are not related to cost/benefit analysis in local authorities but are largely dependent on the personal perceptions of the respondents according to her or his every day experience in the authority and to the personal knowledge about the technology.

Table 8.7 shows the main group of benefits, including improved information processing, better quality decisions and savings . The other factor included issues not identified by the questionnaire. Similarly to the British survey (Campbell and Masser, 1992) it is striking that the vast majority (66%) regarded improved information processing facilities as the most important benefit related to the use of GIS; 23% regarded savings as the main benefit and only 2% identified better quality decisions as most important. The Danish AM/FM and GIS applications are primarily used for mapping and managerial purposes. This may explain why better quality decisions were ranked very low. There might also be a close connection between the priority of achieving a reduction in time spent on improved information processing and savings in relation to the clear political signals in most authorities about a reduction in staff and expenditure.

Out of 33 respondents, 11 (33%) highlighted technical problems as the most serious group of problems while data related and organizational problems came equal second

Table 8.7 Most important groups of benefits[a]

	Improved information processing		Better quality decision		Savings		Other		No reply	
	No.	%	No.	%	No.	%	No.	%	No.	%
Municipality										
>100 000	1	33	0	0	1	33	1	33	0	0
50–100 000	7	78	0	0	2	22	0	0	0	0
25–50 000	16	70	1	4	3	13	0	0	3	13
10–25 000	28	61	1	2	11	24	1	2	5	11
<10 000	33	69	0	0	13	27	1	27	1	2
Total municipalities (n = 129)	85	66	2	2	30	25	3	2	9	7
Counties (n = 9)	2	22	1	11	1	11	0	0	5	56

[a] Has been answered by local authorities which had GIS in January 1993 or have firm plans to adopt GIS in 1993–94.

Table 8.8 Most serious groups of problems[a]

	Data related		Technical		Organizational		Educational		Other	
	No.	%	No.	%	No.	%	No.	%	No.	%
Municipality										
>100 000	0	0	0	0	0	0	0	0	2	100
50–100 000	1	17	1	17	3	50	1	17	0	0
25–50 000	3	30	4	40	2	20	1	10	0	0
10–25 000	3	23	5	38	2	15	1	8	2	15
<10 000	0	0	1	50	0	0	0	0	1	50
Total municipalities (n = 33)	7	21	11	33	7	21	3	9	5	15
Counties (n = 6)	3	50	0	0	1	17	0	0	2	33

[a] Has been answered by local authorities with a GIS in January 1993.

(21%). Table 8.8 provides a breakdown of the findings which demonstrate that the small municipalities are facing more severe technical problems. The larger municipalities place more stress on organizational problems while in the medium sized municipalities there is no clear pattern.

A GENERALIZED MAP OF ADOPTION AND USE OF INFORMATION TECHNOLOGY
IN LOCAL AUTHORITIES

The findings presented in the previous sections indicate that digital register information systems, AM/FM, and GIS do not exclude each other but to a large extent complement each other by function. However, they also show that so far the integration between 'old' information systems (register based systems and mapping systems) and GIS are very limited, and that the strategies of AM/FM and GIS implementation rely almost entirely on a departmental approach. The adoption of AM/FM and GIS has been most pronounced in the larger municipalities and the counties with the most popular equipment being Intergraph Microstation (MGE) or Dangraf on workstations. The small municipalities are more focused on PC solutions starting with an AM/FM system based on cheaper GIS or CAD applications. The present use of the GIS and the AM/FM in all types and in all sizes of authorities is limited to the Technical Department, where units for mapping and the utilities are the primary users of the GIS or the AM/FM applications. Units responsible for local planning, building permission, strategic planning and environmental planning may have access to the systems for hard copy printing of maps or for simple queries.

The majority of the municipalities have stressed improved information processing as the most important group of benefits and the most serious group of problems is connected with the technical and organizational context.

Figure 8.7 is a 'snapshot' of the implementation and the use of GIS technology as well as register based information systems in the last 10 years and describe 'the coherent digital space' and the 'ferry links' and the 'bridges' between the different systems and applications. It draws on the generalized findings from the survey and the historical process of GIS implementation in the Danish context. This 1992 map displays two 'continents': the 'management continent' and the 'map continent'. The first continent is composed by the units of tax, budget and social care. The island of 'health care' is located in the counties. The use of IT is—apart from word-processing—dominated by register based information systems and management systems many of which are based on mainframe technology and the use of on-line terminals. This 'continent' was originally a number of smaller 'islands' but during the 1980s a common data set related to management of the local authority has formed a more 'coherent digital space'.

The 'map continent' is much smaller and has been established recently. It is composed of the units of mapping and utilities. Close to this 'continent' there are a number of small 'islands': building permission, local planning, road and transport (Figure 8.7). The use of IT is dominated by the use of PCs for word processing but a common set of digital map data has established a wider 'digital space'. The communication between the 'map-continent' and the 'islands' is limited to physical transport of hard copies or digital copies of maps. The two 'continents' are located far from each other and have almost no 'ferry links' connected.

The future development trends of this 'IT landscape' can be identified in the survey findings: some of the 'islands' (building permission, local planning and road) near the 'map-continent' will be more closely linked to it, and the 'ferry links' to strategic planning, environmental planning will be improved. However, there is no evidence in the findings that a development towards closer links between the two 'continents' will take place.

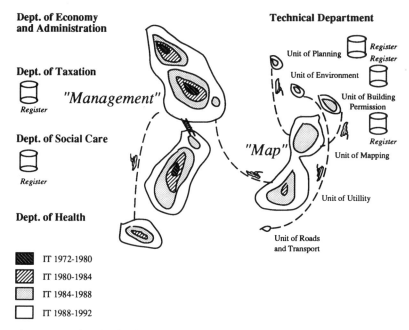

Dept. of Economy and Administration

Technical Department

Unit of Planning — *Register* *Register*

Unit of Environment

Dept. of Taxation

"Management"

Register

Unit of Building Permission

Dept. of Social Care

Register

"Map"

Register

Unit of Mapping

Unit of Utillity

Dept. of Health

Unit of Roads and Transport

IT 1972-1980

IT 1980-1984

IT 1984-1988

IT 1988-1992

Figure 8.7   The islands of information technology in Danish local government.

CONCLUSIONS: TOWARDS A TYPOLOGY OF GIS DIFFUSION IN DENMARK

The analysis of the information provided by the survey indicates that there is a close connection between the differences in size and economic capacity of local government and differences in the implementation strategies adopted to GIS and other information technologies. This is one of the major findings which emerged from the analysis. The distinct nature and costs of different systems have a big influence on the decision process and management of the organizational and temporal tasks in the implementation process. The consequences of the different strategies vary considerably. On the basis of the results the following four-fold typology of implementation strategies has been developed for large, medium and small municipalities and counties respectively:

Large municipalities (over 50 000 inhabitants, n = 14):   All municipalities in this group have adopted GIS or an AM/FM system today and most of the systems were implemented in the late 1980s; 55% of these are 'real GIS'. The main applications have been implemented in the areas of digital mapping and the management of pipes and technical infrastructure in the utility departments. Units related to local planning, building permission and strategic planning have access to the digital maps and they are mostly using them for printing hard copy maps or for simple queries about the use of land or the use of buildings. Generally the systems are not linked to any big register information in Kommune Data. Some of the 'older' systems run on mainframes but most municipalities have preferred workstations as a hardware platform. The software adopted is Dangraf, GeoCAD, Computer Vision and Microstation MGE of which the first was developed by Kommune Data (no longer developed but still supported).

Some 30–40% of the big municipalities have firm plans to implement GIS in 1993–4 and 60% of municipalities with AM/FM today want to change to a 'real GIS' within that period of time. It is most likely that a future strategy of implementation will focus on the use of GIS in units of the technical department with the focus on distributed systems run on workstations in network. The departments of education and social affairs have no plans to adopt GIS and will continue to use 'old' on line, register based information systems for some years. The respondents in the big municipalities have stated that organizational problems are the most serious group of problems related to the implementation process and as in all other municipalities they consider easier access to information and improved information processing as the most important group of benefits related to the use of GIS.

Medium size municipalities (10 000–50 000 inhabitants, n = 123)   Some 62% of the municipalities in this group have adopted GIS or AM/FM systems, most of them within the last 2 or 3 years. Only 18% of these systems are 'real GIS'. The main area of use is mapping. Some of them are handling digital maps in the units of utilities, local planning, building permission and strategic land use planning. This use is close to the pattern of the big municipalities though there seems to be a slightly broader use of these systems in this group of smaller and middle size municipalities. In some cases the units for property management and the departments of education and social affairs have access to GIS as well as the units responsible for road construction and environmental planning in the technical department. However, this is more a case of distributed systems without direct links to registers. Most systems run on workstations or PCs in a network or as a stand alone. The dominant software packages are Dangraf, GeoCAD, Microstation (MGE), and AutoCAD (+ Point or LIGS). But also smaller PC based systems like MapInfo are represented.

Some 30–40% of all the municipalities in this group (same level as *big municipalities* and *small municipalities*) have firm plans for implementing GIS in 1993–4 and 50% of municipalities with AM/FM today want to change to 'real GIS' within that period of time. It is most likely that a future strategy of implementation will focus on the use of GIS for AM/FM functions especially in the units for utilities and land management and to some extent in the units for strategic planning. Some municipalities use 'old' register based information systems (Kommune Data) and smaller DTP packages for map presentation of register information, in parallel with GIS and AM/FM. Though the survey has identified some minor use of GIS outside the technical department it is most likely that the department of education and the department of social affairs will use 'old' on-line, register based information systems for some years. The most serious group of problems is related to technical issues.

Small municipalities (under 10 000 inhabitants, n = 138)   The majority of small municipalities have not adopted GIS or AM/FM. Only 25% have adopted AM/FM systems, most of them within the last 2 years. Only two out of 138 have implemented GIS. Many AM/FM systems are based on CAD technology and run on PCs, mostly as stand alone systems. The AM/FM systems are used only in the technical department for the handling of the digital maps for tasks related to utility, sewer and local planning. Some 30–40% of the small municipalities have firm plans to purchase GIS in 1993–94. They are most likely to be PC based systems used for utility and land management and Desk Top GIS for map presentation of a core set of register information. In the technical department some municipalities will still use manual paper based files or 'old' on line or distributed information systems for land management, building permission and environmental planning. The departments of education and social affairs are likely to keep on using the 'old' register based information systems provided by Kommune Data.

Counties   Due to the small number of counties it is difficult to present a clear picture of the implementation process here; 63% of the counties have GIS or an AM/FM most of which were adopted within the last couple of years. The implementation strategy seems to be dominated by a departmental approach and apart from handling the digital maps the use of the systems seems to be limited to more narrow tasks in the technical units as the unit of rural planning, the unit of environmental planning and in a couple of cases also the unit of public transport and the unit of road maintenance. Two systems are dominant in the counties. These are Microstation (MGE)/Microstation (CAD) and MapInfo running on workstations or PCs. Only one county has both systems. It is likely that these systems will dominate the future implementation strategies in the counties. There are no signs of diffusion of GIS to the departments of health care or social care.

REFERENCES

ANDERSSON, U., 1991. *GIS användning i Sverige*, ULI rapport 1991:3, Sweden.

BATLEY, R. and STOKER, G. (Ed.), 1991. *Local Government in Europe—Trends and Developments*, Basingstoke: Macmillan Education Ltd.

BOLIGMINISTERIET, 1986. *Behovet for et landinformationssystem*, Copenhagen.

BRANDE-LAVRIDSEN, H., 1989. Digitale tekniske kort, standarder og formater, *Landinspektøren*, **34**(6), Copenhagen.

CAMPBELL, H., 1993. GIS implementation in British local government, in Masser, I and Onsrud, H. (Eds) *Diffusion and Use of Geographic Information Technologies*, Dordrecht: Kluwer.

CAMPBELL, H. and MASSER, I., 1992. GIS in local government: some findings from Great Britain, *International Journal of Geographical Information Systems*, **6**(6), 529–46.

LARSEN, K.H., 1989. Matrikeldirektoratet- et tilbageblik, *Landinspektøren*, **34**(6), Feb., 1989, Copenhagen.

MATHIESEN, K., 1983. *Local Government in Denmark*, Copenhagen: The National Association of Local Authorities in Denmark.

MOURITZEN, P.E. and NIELSEN, K.H., 1988. *Handbook of Comparative Urban Fiscal Data*, Odense: DDA.

NIELSEN, O., 1991. Key issues in the local government debate in Denmark, in Batley, R. and Stoker, G. (Ed.) *Local Government in Europe—Trends and Developments*, Basingstoke: Macmillan Education Ltd.

TROLLEGÅRD, S., 1992. *Information Management Policy by Co-ordinated Information Systems*, Copenhagen: Department of Housing.

PART III

GIS in Local Government in Four Other European Countries

CHAPTER 9

Greece: the development of a
GIS community

DIMITRIS ASSIMAKOPOULOS

INTRODUCTION

This chapter analyses the diffusion of GIS in local government in Greece not from
a quantitative perspective, as the diffusion process has only started recently in this
country but from the point of view of the role of informal networks of key individuals
fostering the diffusion process. This is made possible by the small size of the country
(10 million people in 1991) which allows the identification of the small core of
individuals at the heart of the rapidly developing Greek GIS community.

This community includes some 60 key people working in different organizational
and administrative settings throughout the country (Assimakopoulos, 1993). Half of
them are based in university departments, with the rest divided between public sector
organizations (30%) and private sector firms (20%). In terms of professional
backgrounds, half are surveying engineers, a quarter are architects/planners and the
remainder come from backgrounds such as civil engineering, information technology,
geography, geology and agriculture. As an administrative layer, municipalities
constitute one of the main groups of users, where one can explore the diffusion of
GIS in Greece at the very early stages of the adoption and implementation of such
technological innovations.

With these considerations in mind the chapter is divided into five sections. The
first two describe the institutional context of local government in Greece and discuss
the role of external factors such as EU funding and emerging social and communication
networks in Greece or abroad. The third section provides an overview of GIS
diffusion in Greek municipalities with two case studies in the municipal authorities
of Volos and Kos. Then the fourth section evaluates the current situation in terms of
factors and processes which are likely to stimulate or inhibit the further diffusion
of GIS in Greek local authorities, while the final section draws the main conclusions
and highlights further research areas in terms of GIS diffusion in Greece.

LOCAL GOVERNMENT IN GREECE

Structure and size of Greek Local Authorities

The system of Greek local government consists of two different levels, regional and local. Regional government is controlled by central government and includes two types of administrative units: 'periferia' or regions, and 'nomos' or prefectures. Local government is directly elected by the local population and also consist of two types of authorities: the 'dimos' or municipality, and the 'koinotita' or village (see Figure 9.1). Only ten out of the 361 Greek municipalities have a population more than 100000, while 304 have a population of less than 30000 (Velonias, 1991). However it should be noted that two thirds of the total Greek population (10 million) lives in urban areas and the other third in villages. Almost half the population live in the two biggest cities: Athens (3.4 million) and Thessaloniki (0.9 million).

There are 23 out of the 51 prefectures with a population of less than 100000 and another 24 have populations between 100000 and 250000 [calculations based on Velonias (1991) pp. 149–54]. The 13 periferia (or NUTS III in terms of Eurostat) were restructured in 1986 towards EU integration in 1992 and as newly established regional authorities cannot compete with the prefectures which have existed for more than 50 years. The prefecture as a governmental institution plays a significant role at the local level as channel of funding and control from central government.

Functions and Resources of Greek Municipalities

In terms of functions which are related to geographic information the municipalities have the following main responsibilities: (a) to design, construct, maintain and manage water and sewerage networks under the newly formed and financially independent municipal utility companies, (b) to develop and maintain municipal roads, squares and bridges, (c) to manage the municipal bus network, although in many Greek cities this now belongs to the private sector, (d) to organize and manage refuse collection, (e) to design, construct, maintain, and manage markets, parking, sports grounds, libraries, cemeteries and any other municipal buildings or assets. Some municipalities are responsible in their topographic and planning departments

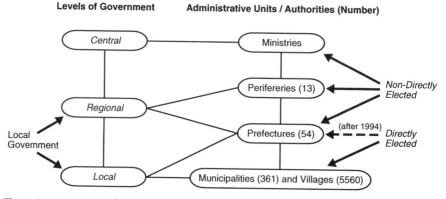

Figure 9.1 Structure of local government in Greece.

for planning permissions but in many of the Greek cities this service is directly delivered by a central government office which is under the Ministry of Environment, Planning and Public Works and is supervised by the local prefecture.

In terms of resources Greek municipalities are very poor compared with their European counterparts and suffer like the rest of the public sector by severe lack of money. It should also be noted that most Greek municipalities do not yet have IT services in-house and they sub-contract the development of various computerized applications for their payroll, their birth, marriage, and death records, as well as for tax purposes to private sector firms. In addition, during the last 2–3 years the municipalities have not employed new staff including specialists who are needed for the introduction of technological innovations like GIS, because of their financial shortcomings and because of a wider governmental policy of reducing the staff in the public sector.

Greek municipalities are also very poor in terms of the availability and access to digital topographic and socioeconomic data. In Greece there is no cadastre (Rokos, 1989) and up to now each municipality has its own policy for data (topographic, socioeconomic, land-use, etc.) collection, updating, and integration. The Hellenic Military Geographical Service (HMGS) which is responsible for the production of maps from 1:5 000 to 1:1 000 000 scales, has released and sells in digital form only the contours at every 20 m for the Athens and Thessaloniki areas from its 1:5 000 scale topographic map, and a national road and administrative boundaries map at 1:1 000 000 scale (HMGS, 1992). This situation can be explained from two viewpoints. First, HMGS is under the Ministry of Defence, so the military decision makers are not concerned with GIS in local government. Secondly even if the military decision makers wished to produce this kind of digital topographic data for civilian uses they do not have the financial resources to do so.

However it should be noted that there is an increasing number of private GIS consulting firms that are becoming digital data producers. For example, InfoCharta produced a digital street map (similar to TIGER) for the metropolitan area of Athens in 1991 which has been marketed by Omas (the Greek MapInfo vendor); and Eratosthenis produced for Eurostat in 1992 the administrative boundaries for the Greek municipalities and villages using the 1981 census population data (Doganis, 1992). On the other hand, although the National Statistical Service (ESYE) had sub-contracted the Institute of Applied & Computational Mathematics in the Research Center of Crete at Heraklion to develop a PC based application for the socioeconomic data of 1981 census (Prastacos *et al.*, 1988), it chose not to disseminate this product. Up to now ESYE releases and sells the socioeconomic data derived from the Greek census only in a hardcopy form.

EXTERNAL FACTORS

EU Funding

The role of EU funds is crucial for the GIS field in Greece, as a major part of GIS investment to date in all administrative settings comes from various EU programmes. The Mediterranean Integrated Programme funded the adoption of GIS innovations in the National Mapping and Cadastre Organization, the National Institute of Geological and Mineral Research, as well as in the Directorate of Forests at the

Ministry of Agriculture. Furthermore, specific programmes of the various EU Directorates have substantially funded major GIS efforts such as the URSA-NET project under the Comett programme. The CORINE programme has also developed environmental GIS applications on a national scale and the ENVIREC programme specifically funded ten environmental GIS projects in 1992–93. The DRIVE programme also funds the development of a GIS application in the City of Piraeus for the traffic management of its harbour.

In terms of Greek local government the EU Social Fund (EUSF) has played a key role in the adoption and implementation of GIS in approximately 30–40 Greek municipalities. For the great majority the main reason to adopt GIS software and hardware was that more than two-thirds of the financial costs were covered by EUSF. The EUSF sponsored in the late 1980s and early 1990s various seminars for the training of personnel in Greek public sector organizations in the areas of new technologies including GIS in municipal authorities. Therefore various private sector consulting firms and university laboratories organized GIS seminars offering in their package not only GIS education and training but also a PC 386 or 486 plus a software package, most of the time a PC Arc/Info. Nikea and Ano Liosia in Athens, Kalamaria in Thessaloniki, Kalamata, Rodhos, Kos, Volos, Patras, Thiva, Lamia, Piraeus, Heraklion, are a few of the City Councils where a start has been made to develop GIS applications under this scheme (see Figure 9.2).

Figure 9.2   Location of case studies.

Communication Networks in Greece and Abroad

GIS events like seminars, forums and conferences in Greece or abroad (Europe or USA) have played a crucial role for the transfer of GIS to Greek local government. GIS are still developing very fast so given the pace of current progress the state of knowledge is fragmented and personal contacts play a major role in the transfer of GIS technology. Moreover because GIS are an inter-disciplinary technology which crosses the boundaries of many well established scientific and technological communities the participation of people in various GIS events creates the necessary links among people from different academic and professional backgrounds, or even different countries, who share a common interest in terms of GIS adoption, implementation and use.

Rogers and Kincaid (1981) claim that a network approach is a method of analytic and strategic thinking in modern societies which is 'knowing from whom to obtain information in order to make important decisions, becomes a critical quality for individual (and organizational) effectiveness in today's society' (p. 344). For example, in Greece there are certain people who are linked with the international GIS community as well as expatriate Greeks who live and work in the EU or the USA and have connections with their homeland. Most of the Greek GIS experts have attended seminars and conferences in Greece but very few have participated in conferences abroad. The most popular GIS event in Greece to date is the Arc/Info ESRI Users' Conference which is organized by Marathon Data Systems in Athens. More than 300 people throughout Greece who are interested in GIS technology attended each of the conferences organized between 1991 and 1993.

The first Greek conference and seminar about GIS technology was an advanced seminar about computers in planning organized in Patras, in 1988, by URSA-NET (Urban and Regional Spatial Analysis Network for Education and Training), a network of Universities, Enterprises, and Local authorities under the EC COMETT programme (Polydorides, 1993). GIS were included as a key area of interest and experts from the Netherlands, England, France, Denmark and Italy, presented papers about their experiences. Since then URSA-NET has organized six such conferences/ seminars also presenting the experiences of Greek experts and Greeks who work with GIS in the USA and the EU countries. The latest URSA-NET training seminar, Forum '92, was organized in June 1992, at the Laboratory of Urban and Regional Planning, University of Patras, and 30 local authority officials from eight different municipalities across Greece attended. In this sense, the URSA-NET project opened a communication channel at an EU level providing the opportunity for European GIS experts to meet and discuss GIS technology with Greek GIS experts.

GIS IN GREEK MUNICIPALITIES

The institutions that are already active in Greece developing various GIS applications are central government organizations and the municipalities. The applications vary from general map production and thematic mapping to network management, and from environmental monitoring to military applications at the national level. Future GIS applications will be developed mainly by the utilities (telephone, water/sewerage, etc.) for the production and updating of their digital maps and the management of their networks. As this chapter is focusing on the local government level the next section will concentrate on the efforts of various municipalities. A broad overview of GIS applications together with a discussion about the benefits and problems for GIS technology in Greece can be also found in Assimakopoulos (1993).

Approximately 50 Greek municipalities have adopted GIS software and hardware during the last 5 years. The PC and Unix versions of Arc/Info have become the predominant GIS software representing around 80% of the total installations. After installing the software some municipalities have also started to develop GIS applications (see Figure 9.2) but the results in most of these cases are limited to pilot projects. Up to now the main applications have been related to the production of updated digital topographic base maps at 1:1000 or 1:2000 scales. The water/sewerage organizations which are financially independent, but most often work together with the technical services of the City Councils, want to adopt and implement GIS as well in the next 2–3 years for the production and updating of digital maps of their networks since in most of the cases the paper maps do not exist at all.

The importance of external funding for the diffusion of GIS in Greece is indicated by the fact that the first application of GIS in a Greek municipality was in the city of Kalamata after the 1986 earthquake as a part of the reconstruction programmes of the City Council through EU aid. Similarly, Rodhos City Council has been developing a GIS application since 1988 for the restoration of the medieval part of the city with the cooperation of the Committee of Byzantine Antiquities and the support of the United Nations Environmental Programme. Nikea in the Athens area is continuing at the present time an interesting GIS application which started in 1989 with the support of the EUSF. A team of 21 young unemployed surveying engineers worked for 2 years with the main objective of getting familiarized with GIS technology and developing various applications, such as thematic maps for land-use planning, and transportation management for the City Council. As a result of their efforts the municipality has a digital base map at 1:5000 scale for all the city area and many layers of thematic information which are managed by 2–3 people from the original team (Tsivili and Domitsoglou, 1991). The municipality of Ano Liosia, also in the Athens area, started an application at a very early stage in the diffusion process of GIS in the Greek local government in 1990, but at the present time it seems that the project has been abandoned. The Laboratory of Cadastre, Photo-grammetric Engineering and Remote Sensing, at the Department of Rural and Surveying Engineering, Aristotle University of Thessaloniki (AUT), produced various thematic maps using GIS for Kalamaria City Council in the Thessaloniki area (Livieratos et al., 1990). The same academic research team is developing GIS applications in Volos and Lamia City Councils.

The Laboratory of Urban and Regional Planning, at the Department of Civil Engineering, University of Patras, developed the Patras Urban Information System (Polydorides, 1992). This first phase of the project finished in December 1992, and at the present time (August 1993) the project faces many organizational difficulties which stem from the administrative structure of the City Council's technical services. The same academic research team is now developing GIS applications for the City Council of Thiva. The prospects seem more promising because the whole effort started from inside, the local authority adopting a bottom-up approach and the size of the city is much smaller than Patras.

The three biggest Greek City Councils (Athens, Thessaloniki and Piraeus) have recently started developing some GIS applications. Eratosthenis is developing an application for the traffic management of Piraeus harbour which is one of the biggest in the Mediterranean, and recently started two GIS projects with the Piraeus development corporation which includes all the City Councils in Piraeus metropolitan area, related to the cadastre and urban planning, and with Athens City Council cleansing department for refuse collection. Thessaloniki City Council is developing a cadastral GIS application in house in its topographic department and with the help of its IT services. Eratosthenis has also been developing various GIS applications for the city of Kos since 1990 as illustrated later in this chapter.

During the next 1–2 years many more municipalities seem likely to invest in GIS with a view to developing applications for cadastral, facility and parcel management, and tax assessment. The latter is quite important as there has been no property taxation to date in Greece. A new property tax was introduced by the central government on top of existing taxes in 1992 to fund the municipalities, removing the financial burden from the government. Although this new tax does not require the cross referencing of surface area data provided by the national electrical company with other data based on existing topographic maps it provides a justification for the development of municipal GIS applications. To date every owner of a parcel must declare to the national electric company the surface area of the house or his/her property before this property is connected to the electricity supply. These data were used for the implementation of the new property tax in 1993, but there are some City Councils which want to cross-reference this information using topographic or other methods, collecting primary data for the production of an urban cadastre in the long run. In the short run this exercise could raise more income for the municipal authorities feeding back to the development of a municipal GIS.

After this brief overview of municipal GIS in Greece, it is time to present the findings from the two case studies in the cities of Volos and Kos.

Two Case Studies

The pioneering City Councils at Volos and Kos have very recently established separate GIS Departments under their Technical Services with the objective of developing GIS applications at an operational level. Until 1990 both municipal authorities had very little experience of microcomputers or other information technology and management systems. For example in many departments (e.g. water/sewerage) even paper-based record-keeping did not exist at all. As Calhoun et al. (1987) argue in their case study of IT implementation in a system-poor environment, at the present time in these two Greek municipal authorities 'microcomputers and workstations are perceived and managed as if they were mainframe computers' (p. 369) mainly because of the unwillingness of the end-users to accept computer use as a part of their everyday jobs. At the present time, users perceive computer use as an apparent increase in responsibility which is not counter balanced with any kind of tangible reward. The background, the current situation and the future prospects of each of these projects, together with some technical and organizational issues are discussed below.

Volos municipal GIS

Volos is the capital city of Nomos (prefecture) Magnisias and one of the medium to large municipalities in Greece with a population of 80 000. Volos lies halfway along the motorway between Athens and Thessaloniki (see Figure 9.2), and as the fourth biggest Greek port connects Greece, as well as other EU countries, with the Middle East (Syria).

The GIS project at Volos City Council started in early summer 1991 when an employee of a financially independent municipal company (DEMEKAB) which undertakes the design and construction of various municipal projects, realized that it was possible to fund a GIS project through the European Union Social Fund. He discussed the issue with a surveying engineer in the Planning Department of the City Council who decided to organise a 2-day GIS workshop inviting the academic research team based at the Department of Rural and Surveying Engineering at the Aristotle University of Thessaloniki (AUT) to demonstrate the potential of GIS/LIS technologies for Greek municipalities.

The surveying engineering academic research team based at the AUT presented various potential applications including an automated municipal atlas which had been developed earlier for the City Council of Kalamaria in the Thessaloniki area (Livieratos et al., 1990) at a 2-day workshop in June 1991. A few weeks later a 2-year research contract was awarded by the Municipal Company of Design and Constructions (DEMEKAB) to the university team to develop a pilot project for a municipal GIS.

At a higher level the vice-Mayor who was responsible for the municipal technical services including the Planning Department supported the project from the beginning, since he realized that it was a rather inexpensive way to introduce high technology and modernize many of the existing paper based services, as well as developing some new products or services. It should be noted that various existing municipal paper maps were significantly outdated and some of those printed in the 1930s are in a very bad physical condition. The Planning Department of Volos City Council was also responsible for granting or refusing planning permissions since 1981 and had organized a paper based archive with plans and documents with almost every new building in the city area from 1981 onwards. It was generally expected that this archive would become a part of the municipal GIS project and the information would be stored, analysed, and displayed in a more sophisticated way, assisting the city planners and many other users in their everyday tasks. These kinds of needs and expectations deriving from the planning and other Departments justified the initial investment, in terms of time, money, and effort, so a broad consensus about the project existed at all levels of potential users.

Overall three interrelated factors played a major role for the set up of the municipal GIS at Volos. First the existence of two interconnected champions in strategic positions in the municipal authority. The DEMEKAB employee can be perceived as a champion who undertook the responsibility for the financial support of the GIS innovations, while the surveying engineer at the Planning Department was a technical champion who undertook the responsibility for the development of the GIS applications at the technical services of the City Council. Secondly the existence of a financially independent municipal company (DEMEKAB) to direct and manage the EU financial aid with the political support and backing of the vice-Mayor responsible for municipal Technical Services. Thirdly the fact that the surveying engineer had been in the Planning Department since 1987 and was to a large extent familiar with the use of PCs for the production of digital topographic maps. During the late 1980s he had also worked for his own small technical bureau on the production of digital topographic maps using AutoCad software in a PC environment under a planning effort (EPA) of the Greek Ministry of Environment, Planning and Public Works. Consequently he had a personal interest to learn about GIS.

The AUT research team adopted a two stage strategy for the introduction of GIS in the municipality of Volos (Boutoura, 1992). In the first stage from 1991 to 1993 an introductory GIS application was developed at AUT using Intergraph's Micro-station and dBase III plus software on a 386 PC. This project finished in June 1993 and is called DIMOS (dynamic information municipality oriented system). A team of four academics was involved in its development, all of them surveying engineers, from the Laboratory of Cadastre, Photogrammetry and Cartography. In its second stage, in the next few years, DIMOS-VIP (Volos Integrated Package) may be developed in a Unix environment using Intergraph's workstations and MGE software. According to Boutoura (1992) 'Our main concern was to develop something simple and user friendly. The total absence of previous experience among the users was for us a strategic criterion for the design of the system, which had to be fully interactive' (p. 1.24)

Nevertheless the GIS innovation implementation efforts at Volos faced a lot of difficulties not only because of the lack of technical skills and the lack of digital data but most importantly because of political and organizational issues. In terms of data, the DIMOS project used two topographic base map sheets at 1:1000 scale covering part of the city centre. Because both of these base maps were last updated in 1982 the AUT team had to work in the field and update the maps as well as collecting primary information for the development of their databases. The spatial unit of analysis for DIMOS is the building, the land parcel, or the urban block, and the descriptive information concerning the development of its applications includes socio-economic, land-use, transportation, and other data-sets which to a large extent was collected in the field. The research team also used aerial photography of part of the city and a photo of a building in order to incorporate them into DIMOS.

In March 1992 a number of thematic maps at 1:5000 scale and a prototype of a municipal atlas for Volos were produced at AUT for this small area of the city centre. Since then a digital base map at 1:1000 scale has been produced by the municipal GIS department covering the whole city area. This map includes 45 paper map sheets which were produced in 1982 from aerial photography, and need updating over the next few years. A great deal of money, time and effort is needed to collect and convert in a digital form the municipal GIS and other departments primary and secondary socio-economic, land-use, and other data for the development of various GIS applications.

Despite the initial successes the overall development of the project changed dramatically in the first half of 1993. In March a new vice-Mayor took over the responsibility for the technical services. Under a new municipal internal regulation in early summer 1993 the new GIS Department was officially established together with its formal positions. As a result the original technical champion is now limited to his job in the Planning Department while the new formal position for the head of the GIS Department has been given to a junior member of the Planning

Department staff, also a surveying engineer, graduate from AUT in 1989. In late July 1993 the GIS Department was in a transition stage and the situation was unclear in terms of both leadership and future prospects.

Kos municipal GIS

Kos is a small municipality of 14 000 people with a totally different profile from Volos. The island of Kos is located in the far east of the Greek state only 15 km away from the Turkish border (see Figure 9.2). It is one of the provinces of Nomos Dodekanison and together with Rodhos island is one of the more developed and cosmopolitan tourist resorts in Greece. The greatest part of its growing economy is based on the tourist industry. All 12 islands of Nomos Dodekanison including Kos were under Italian (Venetian) rule, but after the Second World War, in 1948, they became part of the Greek state.

From 1925 to 1935 the Italians organized a Cadastre Office in Kos which covers all the island of Kos and part of Leros at scales 1:1000 for the urban areas and 1:2000 for the rural areas. Since 1948, this Cadastre Office has operated under the Greek Ministry of Justice, storing and updating paper records for every parcel or building in a way that all the history of ownership and geometry is shown through maps, diagrams and books. Today the volume of information which has been stored is too large to be managed manually while the physical condition of many maps and diagrams is deteriorating. For example there are 334 books with 200 pages each covering a total of 30 000 properties and there are 239 cadastral maps/sheets which have not been updated since their original design because the history of geometry of every parcel is stored in a file on a case by case basis and not on the map (Takas and Veinoglou, 1990). Despite the arguments put forward by Kos City Council, that if all this information was properly managed it could substantially help the development and planning of the municipality and by and large of all the island, the City Council has not succeeded in incorporating the Cadastre Office in its authority.

The municipal GIS project in Kos officially started in late 1990. From 1983 to 1987, Eratosthenis, a consulting firm based in Athens, was employed under the EPA programme of the Greek Ministry of Environment, Planning and Public Works to survey 200 ha of suburban land to produce a proposal for the extension of the Kos City plan. The output of this project was in digital form including a digital topographic base map at 1:1000 scale which was produced using AutoCad. But according to Siorris *et al.* (1989) 'the main problem was the correlation of the current situation as was measured on the ground with the information that was recorded at the Cadastre Office'. As the Cadastre Office is an independent authority there was a need for formal cooperation with Kos City Council at many levels involving various legislative, technical and financial issues. In May 1989 a study by Eratosthenis on behalf of the Kos City Council showed how a new technology like GIS could help the implementation of the extension plan automating the whole process, preparing also for the future when a municipal cadastre office could be set up independently, or in cooperation with the existing one (Takas and Veinoglou, 1990).

Kos City Council as a whole appears to be commited to technological innovation. For example the municipal TV and radio stations have covered the issues surrounding the introduction of IT and in particular GIS in the municipal technical services. The mayor of Kos is extremely popular among the local people. He has been mayor for 17 years and was re-elected with 70% of the vote in 1990. The Planning Department of the municipal technical services had also prepared the ground for GIS through a number of studies in the 1980s concerning the extension of the City Plan, the study of different neighbourhoods/areas of the city and the zoning of small scale industry.

In late September 1990 the City Council decided to establish a municipal GIS, taking into account the recent policy of the Ministry of the Presidency of the Government which allowed the purchase of IT in the wider public sector organizations of up to $600 000 per year without needing approval from the central government authorities. This initial effort for the adoption of a municipal GIS was organized by Kos City Council under three contracts: to Eratosthenis for the development of the project, to Marathon Data Systems (ESRI vendor) and Ingres Hellas for the software which included a triple licence for the Unix version of Arc/Info, PC Arc/Info, and the Ingres (RDBMS) in both Unix and DOS versions, and to a local vendor of Hewlett Packard at Kos for the hardware.

In terms of human resources there were no engineers or other technicians with GIS skills in-house. Two GIS training seminars were organized by Eratosthenis in August 1991 and 1992 funded by the EUSF, and 10 municipal engineers and technicians attended. According to the current GIS project manager, a surveying engineer, who was a tutor in the second of these seminars, the very limited IT experience among most of the participants made it necessary to train them as well in basic keyboard skills and the MS-DOS operating system. The equipment was installed in August 1992 and three people were employed for the implementation of the GIS applications; two were civil engineers and the third was a computer technician. Until early 1993 the GIS project had not developed a clear strategy mainly because the municipal team did not have sufficient knowledge to implement the system.

In April 1993 the marriage of a young GIS expert from Eratosthenis to the head of the municipal Topographic Department boosted the project. A municipal GIS Department was also established at the same time and the young GIS expert became the head of this municipal GIS Department. One of the civil engineers moved to the newly established independent municipal company of water and sewerage to develop a link with the GIS Department. The other civil engineer is presently gathering socio-economic and land-use data for the development of a cadastral GIS application for the collection of the new property tax, while the computer technician was mainly getting familiarized with the Arc/Info software in a Unix environment.

Two more applications have been chosen besides the Topographic and Planning Departments. In the near future GIS applications will be developed for the Cleansing Services, mainly for refuse collection, and the Construction Department for resources management. Eratosthenis has undertaken two additional projects for the digitization of the City Plan topographic maps at 1:1000 scale, and the production of a digital map for the municipal water and sewerage networks, with a deadline of December 1993. More significantly, the Mayor of Kos has displayed a personal interest in this new department which could become a resource for other municipalities in the wider area like Rodhos, so the future prospects of the municipal GIS Department in Kos are very good. At the outset the effective adoption and implementation of GIS technological innovation in Kos was supported not only by a number of interrelated factors such as the existing analogue cadastre, and the strong ties between Eratosthenis surveying engineering consulting firm and the City Council, but there was also an element of pure chance.

EVALUATION

In Greece there is an increasing number of municipalities adopting GIS software. It is estimated that there are approximately 40–50 GIS software licenses installed in summer 1993, most of them for PC Arc/Info. EU financial aid, and in particular the EUSF, has played a significant role, providing a financial incentive for the adoption of the initial hardware and software, as well for the initial GIS training and in the best cases for the production of some digital topographic and attribute data. The implementation process of a municipal GIS in Greece up to now has been mainly undertaken by external contractors (a university laboratory or a private consulting engineering firm) because there is no IT or GIS know-how in-house. These external teams of GIS specialists mainly consist of surveying engineers for two reasons. First, in Greece updated topographic base maps do not exist for most areas and secondly, in the 1980s some of the surveying engineers who work for the technical services of various City Councils became familiarized with the use of PCs for the production of digital topographic maps as a result of a planning effort (EPA) of the Greek Ministry of Environment, Planning and Public Works.

In some municipal Technical Services Departments there are surveying engineers or other mid-career professionals who are to a certain extent familiar with the use of microcomputers and aware of the potential of GIS technology who act as 'champions' for the adoption and implementation of this technological innovation. But as the case study of Volos has illustrated this is not enough for the development of GIS applications at an operational level. GIS applications, in addition to an innovative organizational environment need a great deal of human, financial and data resources which are not available in most Greek municipalities. Moreover a consensus about the project should be developed in the vertical and horizontal levels of an organization during the implementation period of GIS. This requires the support of the highest people in the hierarchy for a long period of time (at least for 4–5 years). This is very difficult, since most often City Councillors or politicians in Greece as in other countries tend to have short term priorities.

GIS applications in Greek municipalities tend to have an inter-departmental and inter-disciplinary nature, because GIS are implemented by various departments (topographic, planning, utilities, etc.). Consequently they should incrementally establish a communication process among different professionals like surveying engineers, architects/planners, civil engineers, etc., for the collection and allocation of various resources, like topographic, land-use, socio-economic, and other data, as well as money, equipment, personnel, etc., to facilitate the effective implementation of a municipal GIS. This might be encouraged either by on the job training opportunities or by financial bonuses, or by the personal interest of the highest officials, as for example happened in Kos where the positive response of the mayor was perceived as a positive signal from the different professionals to spend more time and effort and communicate more effectively for the better understanding and implementation of GIS technology.

At the present time a great deal of valuable human, financial and data resources is wasted, as is reflected in the most advanced of the Greek municipalities where after an initial period of 2–3 years of implementation efforts no GIS applications are yet developed at an operational level. So there is a clear need to recognize and address, in a wide range of Greek organizations and not only in municipalities, that the institutional, organizational and human issues are equal or more important than the technical difficulties which arise during the development of GIS applications.

From the standpoint of technological communities, each of the professional groups (surveying engineers, architects, civil engineers and computer scientists) involved in the creation of the Greek GIS community is a separate well defined professional community. All of them share a common interest in terms of GIS innovations. On the other hand each of these communities has its well established tradition of technological practice and culture, so inherently tends to be very conservative in order to preserve power positions or jobs of its members in the process of adoption and implementation of GIS. In this sense there is a key contradiction between the desire to change incrementally while preserving the existing structure and the need for a new community of individuals and organizations which is related to the GIS innovation.

More importantly there is a conflict at a higher level between the academic and business economic interests of most of the individuals or organizations who shape the Greek GIS community and often have a foot in both camps. This is a hybrid situation where the pledges of loyalty in many cases seem to be implicit, informal, highly complicated, and finally co-exist with both the needs to be individuals or organizations open and closed at the same time.

CONCLUSIONS AND FURTHER RESEARCH

The study of GIS diffusion in Greek local government clearly illustrates an environment where there is a lack of geographic information because of both a lack of resources and the low priority given to information in decision making. Consequently such information as exists is idiosyncratic, partial in nature, and of poor quality. It is very often also dependent on external factors such as the EU funding of specific projects, or the extent and nature of technology transfer mechanisms. An important finding of the above analysis is that for effective technology transfer a key role is played by the degree of organizational openness together with the existence (or not) of external links with various social and communication networks at a national or international level in relation to GIS technology. Therefore further GIS diffusion research could focus on the study of patterns of social and communication network links within the Greek GIS community as these kinds of ties can be seen as the routes which different social actors use to share, exchange or mobilize scarce resources.

It is also worth noting how one of the Greek GIS experts summarized the state of the art of GIS technology in Greek local government: 'GIS in Greek local government is like a Rolls Royce which is used to carry bricks'. This is similar to other contexts such as the British local government where Campbell and Masser (1992) have commented that: 'staff in British local government use GIS for simple things, associating such systems with the straightforward activities of improving information processing facilities rather than the more heavily promoted areas of enhancing strategic decision making or realizing savings' (p. 544). In other words, because of the recency of the GIS phenomenon, there has been only a limited impact so far on existing practices irrespective of whether people and organizations adopt and implement the technology in Greece, or in other more computer literate and information rich countries like Britain. Therefore a second area of further GIS diffusion research could be an in-depth investigation over time of the effectiveness of GIS implementation processes in a variety of organizational contexts throughout Greece and not only in local authorities.

ACKNOWLEDGEMENTS

I would like to express my gratitude to the 60 Greek GIS experts who participated in my research about the development of the Greek GIS community in March–April 1992 or/and June–August 1993. Special thanks are also due to Ioannis Polimenidis and Roula Baklatzi from Volos City Council, as well to Despoina Brokou and Joulia Papaeutihiou from Kos City Council who helped me with the two case studies. This paper has profited by careful readings and useful comments supplied by my supervisor Ian Masser, and Heather Campbell at the University of Sheffield, and by Adonis Kontos and Thanos Doganis in Greece. Since October 1992 this research has developed as a PhD thesis sponsored by the Human Capital and Mobility programme of the Commission of the European Communities (CEC). Any views expressed in this paper are those of the author and do not reflect in any way the views of the CEC.

REFERENCES

ASSIMAKOPOULOS, D., 1993. The Greek GIS community, *Proceedings of the Fourth European Conference on Geographical Information Systems*, Genoa, Italy, March 29–April 1, pp. 723–32, Utrecht: EGIS Foundation.

BOUTOURA, C., 1992. LIS models with applications to municipality development, *Proceedings of the First Seminar of the European Land Information Systems* (ELIS '92), pp. 1.11–1.40, Delft.

CALHOUN, C., DRUMMOND, W. and WHITTINGTON, D., 1987. Computerised information management in a system poor environment, *Third World Planning Review*, 9(4), 361–79.

CAMPBELL, H. and MASSER, I., 1992. GIS in local government: some findings from Great Britain, *International Journal of Geographical Information Systems*, 6(6), 529–46.

DOGANIS, T., 1992. Digital map for the Greek municipalities and communes, *Proceedings of the Second Greek ESRI Users' Conference*, Athens, Greece, 7–8 December (in Greek).

HELLENIC MILITARY GEOGRAPHICAL SERVICE, 1992. *Price List for Geographical Products Sale*, October, Athens: Geographical Policy Department.

LIVIERATOS, E., LIONTA, F., BOUTOURA, C., PAPADOPOULOU, M., SIGALAS, I. and TSANAKA, A., 1990. Automated municipal atlas of Kalamaria City Council, *Proceedings of a Seminar on Geographical Atlases*, November 27–28, pp. 255–78, Thessaloniki: Aristotle University of Thessaloniki, Faculty of Rural and Surveying Engineering (in Greek).

POLYDORIDES, N., 1992. GIS in Greece: a review of the state of the art, *Mapping Awareness & GIS in Europe*, 6(7), 16–18.

POLYDORIDES, N., 1993. Technology transfer and training needs: the URSA-NET experience, in Masser, I. and Onsrud, H.J. (Eds) *Diffusion and Use of Geographic Information Technologies*, pp. 307–16, Dordrecht: Kluwer.

PRASTAKOS, P.P., DIAMANTAKIS, M.N. and MANIOUDAKIS, N.K., 1988. Athina: a PC Based Statistical Database, brief report, Heraklion, Research Center of Crete, Institute of Applied and Computational Mathematics, Division of Regional Planning (in Greek).

ROGERS, E.M. and KINCAID, D.L., 1981. *Communication Networks: Toward a New Paradigm for Research*, New York: The Free Press.

ROKOS, D., 1989. Local government and cadastre, paper presented at a Conference on the Cadastre and Greek local government, Technical Chamber of Greece, Athens, 15 March 1989 (in Greek).

SIORRIS, D., TAKAS, B. and LOULAKIS, N., 1989. The computerization of the extension plan of Kos City Council, paper presented at a Conference on the Cadastre and Greek Local Government, Technical Chamber of Greece, Athens, 15 March 1989 (in Greek).

TAKAS, B. and VEINOGLOU, V., 1990. *Preliminary Study for the Computerisation of the Extension Plan of Kos City Council*, February, Kos: Kos City Council (in Greek).

TSIVILI, A. and DOMITSOGLOU, V., 1991. LIS in local government: the experience from the Municipality of Nikea, *Proceedings of the URSA-NET Forum '91*, Patras, June 7–9, pp. 167–76, Athens: URSA-NET (in Greek).

VELONIAS, E., 1991. The current situation in regional and local government, in Mathioudakis, M. and Andronopoulos, B. (Eds) *The Greek State: Organisation and Function*, pp. 147–202, Athens: Sakoulas (in Greek).

CHAPTER 10

France: a historical perspective on GIS diffusion

PHILIPPE MIELLET

INTRODUCTION

As with any methodological innovation, and to an even greater degree with technological innovation, Geographic Information Systems (GIS) have mobilized interest and their diffusion has even been considered as worthy of research. GIS are not a homogenous area of study because of the multitude of agents representing different groups. The speed of diffusion within these groups varies depending on the geographic environment and the specific needs of the groups' activities. For example, the diffusion of GIS has been rapid within groups that manage networks and within local government. A map of French GIS usage today would show 'spatial' disparities that have different consequences according to the agent or to the area concerned.

GIS are part of a technological environment in a phase of complete transformation for which the two convergent fields (information technology and information services) are today among the most dynamic sectors of the economy. The diffusion of GIS is no longer solely driven by technical progress but also by the increasing number of agents involved particularly in local government. As a result a policy of high levels of financial and technical investment for complex GIS applications is not necessarily the approach adopted by all information users. Some local government organizations have favoured these new approaches that are transforming GIS seemlessly from a management tool to a decision support tool in France as elsewhere. This evolution is a significant and seminal change of recent times that could become, given favourable conditions, a significant trend in the GIS of tomorrow.

In order to understand the current situation of the state of GIS diffusion in French local government it is necessary to briefly go over the major stages of digital geographic information diffusion and to run through their implications in today's market. This will provide the backdrop for the overview of GIS use in local government that aims to be not solely quantitative but also qualitative by highlighting both lines of diffusion and current evolution of GIS use. Given the limited resources of this study, and the availability of a number of surveys recently carried out by associations and private consultants, this study is based on secondary sources of information supplemented by a few selected case studies.

THE DIFFUSION OF DIGITAL GEOGRAPHIC INFORMATION IN FRANCE

As in many other western countries the 1970s saw the first trials with graphic information technology. In terms of geographic information the most important event was the creation of the first Urban Data Banks (the BDU) in the newly created Metropolitan Community Councils (Communautés Urbaines). Unfortunately some of these early data banks were not entirely successful due to the technical limitations of the period. The lack of compatible hardware, the absence of data models, the high cost of digitization for localized data acquisition at the time and the lack of suitably trained specialized staff have often been cited as obstacles to the success of BDU projects.

Even if the fruits of the enabling work carried out during this period did not become evident until the 1980s, the major producers and users of localized data started to carry out studies that aimed at implementing data production systems. For example, in 1970 both the national electricity and gas utility companies started to develop system applications that could locate infrastructure features on a base map. Yet it is only 20 years later that a definition of national products has been achieved. Other experiments by large information producers included the Urban Geographic Inventory (the RGU) by the National Statistics and Census Office (INSEE) that subsequently was partially abandoned (see below).

The 1980s were not a period of uniform progress. The early 1980s saw a continuation of the experimentation of the 1970s. The first stirrings of GIS diffusion only made themselves felt in 1988–9. The decade was a turning point during which the foundations for the major projects that can be seen today were put in place. In 1982 and 1983 the first wide-ranging discussions were held within the framework of the National Commission for Geographic Information (Commission Nationale sur l'Information Géographique) known as the Lengange Commission. The work of this commission was then included within the 9th National Plan that consequently drew attention to the geographic information needs at large scales (the scales compatible with the Cadastre). It was recommended that a topographic land ownership plan (the Plan Topo-Foncier) be prepared that combined the precision of the Cadastre with a topographic element that should include contour and spot height data. A two scale plan was suggested: a 1:5000 scale plan for the whole of France and a 1:2000 scale plan of areas deemed to be priority zones for regional structure planning. These priority areas were evaluated at 10% of the national land surface area (Denegre, 1989). In order to evaluate the real needs of data users, an extensive survey was started in 1986. The results of this survey showed that it was necessary to redefine the data products because most respondents, including many local government users, considered the specifications for the 1:2000 scale plan too restrictive.

The job of working out the specifications for the Plan Topo-Foncier fell on the National Council for Geographic Information (the CNIG). The implementation of the Plan depended on the availability of data from a number of sources such as the Topographic Data Base of the National Mapping Agency (the IGN) and the Digital Cadastral Programme (the PCI) of the National Tax Office (the DGI). Currently the Plan Topo-Foncier is only a proposal on which numerous data users have pinned great hopes for the future.

One of the important consequences of the work of the Lengange Commission was the creation of the CNIG by a decree of 26th July 1985 that came into force on the 28th January 1986. The CNIG aims are: 'to contribute by means of its studies, views and proposals to the promotion of geographic information development and to improve geographic information-related techniques taking into account the needs expressed by both public and private geographic information users' (*Journal Officiel*, 1985). Through different committees (such as the committee for the evaluation of the economic and social use of geographic information and the committee for the standardization of exchange formats), the CNIG aims to increase interest in a community of organizations linked to GIS by acting as the central pivot for wide-ranging enabling projects such as the French digital data exchange format, EDIGEO.

The late 1980s saw the arrival of the term GIS within the geographic database vocabulary. The general use of such terminology has been very recent notably in local government where as late as 1987/8 terms such as data management by 'graphic information technology' were still in use. It was during these years that the specifications for the large national digital databases, including those of the IGN and the DGI, were refined. At the same time satellite images erupted upon the scene with the launch of SPOT I (Satellite Pour l'Observation du Terre) in February 1986 by the CNES (National Centre for Space Research). The importance of the availability of SPOT generated data to the development of GIS cannot be underestimated. Through the program developed to evaluate satellite data (such as the program SPOT AVAL) the CNES has promoted several digital data processing and analysis projects that have in turn galvanized the sector into activity.

The 1980s saw two great revolutions in different fields that continue to have a decisive effect on the rapid evolution of GIS use especially in local government:

1 The first revolution, started by 1988/9, was the appearance of standard GIS software offering functionalities far beyond those that had been available until then: and,

2 The second revolution relates to the impact of the Decentralization Acts of 1982. These Acts transferred responsibilities notably in town planning, in structure plan-making and infrastructure management to local government which as a result started looking for information technology solutions to assist its decision-making in these fields. It is clear that these major institutional changes are at the root of the rapid increase in demand for GIS at all levels of local government: urban authorities, départements, and regions.

The 1990s saw the use of GIS and the number of operational applications really take off. The rapid development of computing hardware (such as increased memory storage, speed of calculation, the general availability of colour screens, the first colour printers and scanners) coupled with decreasing costs made access to GIS and cartographic software considerably easier. The arrival of a great quantity of software on the market (about 50 types of software are commercially available in France) increased the potential for GIS. However, there are still many obstacles to the effective diffusion of geographic information processing in France and one of these obstacles is the notably slow development of standardized, digital data bases.

Basic Digitized Data for Local Government

The diffusion of GIS in local government is undeniably related to the availability of digital data sets. The major reference data base projects were started in the early

1980s but only recently the progress made in their implementation has allowed wider GIS diffusion. The reference data bases currently being developed in France are institutional in nature and some of the principal programs are detailed below.

The IGN (National Geographic Institute)

The initial reflection on the implementation of digital database products at the IGN started in 1983. A survey undertaken in 1986 confirmed the interest of potential users for digital products, and created the conditions for subsequent operational develop- ments. It is important to note that the IGN chose to structure its database both graphically and in terms of attribute data. The consequences of this option are far greater than for a simple digitization of paper maps (by using scanning techniques for example). This option requires the development of true data models that are by necessity exceedingly complex given the quantity of information types listed. The advantages of this approach are that the database can then be used by GIS for both management and analysis but its main disadvantage is that the databases are very complicated to set up.

The cartographic database (BDCarto)   this database contains all the information of the 1:100 000 scale map with additional information. In its final form it has nine information themes (road, rail, water and electrical networks, isolated objects, administrative or technical units, land use and relief). It is precise to 10 m and it can be scaled between 1:50 000 and 1:250 000.

The BDCarto is essentially aimed at medium scale applications that might be used by départements for applications such as road management, transport planning, fire-fighting and environmental management (for example the management of catchment areas). At the end of 1992 some 30 départements had acquired the road network database of the BDCarto (Lamy 1992). This database aims at becoming the reference digital base map for most structure planning applications at départemental or regional levels. Its relative precision (to 10 m) makes it compatible with satellite images that are currently an important instrument for regional structure planning. The publicized willingness of the IGN to regularly up-date the databases (a 1-year cycle for the principal network and a 4-year cycle for basic themes) will allow the operational use of these databases to become a reality in the future.

The topographic database (BDTopo)   The BDTopo corresponds to an enriched 1:25 000 scale map. With 11 themes (road and rail networks, energy infrastructure, water features, buildings, vegetation, administrative boundaries, other boundaries, land relief, geographic and toponomic infrastructure), it will be the most complete and precise database available in France. Its precision to within 1 m makes it a useful tool for detailed structure-planning. Each object is captured in three dimensions and thus it is a database with complete spot height data. The up-dating cycle is projected as being 7 years. Currently some 200 people are working on its development.

The problem inherent to this database is the way it has been set up. Faced with the considerable cost of data capture the IGN has digitized data in an order of priority favouring local authorities that have signed a contract to take the data. The specifications for this database make it an extremely powerful tool but they also make it difficult to manage. One sheet of BDTopo (20 km by 20 km) represents 100–150 mega-octets of memory and it takes 1000 sheets to cover the country

(Boursier, 1993). Due to the complexity of the database characteristics the data model is correspondingly complex and it has to be managed by specialists.

This database is primarily aimed at major users of large scale information. In a 1990 survey, the IGN identified a number of application types that would require this database. Amongst these were many applications linked to structure planning and the environment such as impact studies for infrastructure projects, commercial zoning implementation, and the revision of land use plans (the POS) (Lamy, 1992).

Georoute This is a database of road network information for France that includes all roads in urban areas with more than 10 000 inhabitants and it includes road names and municipal boundaries. Information is localized with a precision of 5 m in urban areas and with a precision of 10–20 m in suburban areas. In 1993, the Ile-de-France, Marseille and Lyon were exhaustively covered by the database that also included the primary road network outside of these areas. The completion of this database is projected for 1996. This database has a topographic structure and is particularly aimed at applications that optimize either networks or itineraries.

The digital data of the DGI (the National Tax Office)

In the family of digital data providers, the DGI is one of the major producers and recent decisions with regard to the choice of information technologies can only reinforce its position. In France the DGI is responsible for the Cadastre. The Cadastre is made up of some 560 000 sheets that describe nearly 100 million land parcels (Dumont, 1992). Since 1990, attribute information has been processed by computer. Moreover the DGI already provides (with restrictions on confidentiality) four files of text data (owners of buildings, of other constructions, of properties without constructions as well as roads, streets and place names). However, the major demand today is to have Cadastral plans that are still essentially recorded on paper support in digital format. Thus the DGI has launched a Digital Cadastral Plan (PCI) for the whole of France, which, when completed, is destined to become one of the fundamental reference databases in France.

Faced with the size of the task in hand, the DGI has sought to develop a number of partnerships by defining an operational framework for the development and the up-dating of cadastral records integrated in GIS and acquired by local government organizations. The principle is simple: digitization cannot be done without the agreement of the DGI which then obtains a magnetic copy of the record and remains the data manager with regards to information up-dating. Special conventions have been agreed with the major network managers (the national telecommunications, electricity and gas utilities). This strategy is similar in principle to that of the Ordnance Survey in the UK and means, again in principle, an accelerated acquisition of information.

The PCI should eventually provide essential data sets for the Plan Topo-Foncier and should constitute the reference base layer for all detailed study work. In order to organize this base layer at the national level, the DGI is currently acquiring GIS (using French Apic software) for all of its 314 cadastre offices.

It is obvious that these reference databases once completed will represent a digital information resource without precedent. They will enable new applications to be developed based on detailed geographic objects. However there has been some criticism from user-oriented groups concerning the use of these databases beyond that of the technical management of the information itself. The setting up of the data models for the databases and their cataloguing has taken into consideration the principal potential clients of these products, i.e. the local authorities. When the specifications were defined (in the late 1980s) the agenda of local authorities in the field of geographic information was set by the technical services of those authorities. Thus the databases have been designed essentially to suit traditional data management applications. The problem that will inevitably surface in the years to come relates to the relevance of the choices made by the data producers (with regard to scales, generalization, precision and the themes) for applications oriented more towards thematic analysis of territorial structures.

GIS IN FRENCH LOCAL AUTHORITIES

Until recently the potential of geographic information and GIS diffusion in local authorities has been perceived as linked to the computerization of working methods in an environment that changes slowly. It has been suggested provocatively that 'French local authorities do not seek, with the exception of some rare examples, organizational revolution. Elected members have not been really aware of the information technology potential. Staff have been very slow to take up the information technology tool' (Dupuy, 1992, p. 37).

This illustrates that organizational obstacles to innovation can be so substantial that projects introducing innovation founder on them. This has led to attempts to formalize GIS implementation methods. For the most part these methods have been derived from implementation theory for information technology in general and they are applicable to local authorities. In France, such methods are mostly derivatives of the Meurise method. This approach has demonstrated the importance of structured and complete studies before a project is physically started. By formally showing the complexities of the geographic information structure in a given organizational context these studies have been a point of reflection for the transfer of knowledge from the realm of fundamental research to the real world. Put in a different perspective these studies have often revealed the institutional logjams linked to the system of keeping localized information in local authorities.

A number of failures in the 1970s, linked to the implementation of systems for managing geographic information in local authorities, raised questions about cost-benefit analysis for digital information use. Following work carried out notably in the USA, Didier (1990) produced 'Utilité et valeur de l'information géographique' that was followed by the work of Didier and Bouveyron (1993); these books have sought to formalize this aspect of GIS projects that is of fundamental importance to local authorities. However within this institutional framework the problem is complicated as the underlying objective is not to achieve an immediate financial profitability but to achieve a social utility which remains difficult to define.

In the context of Europe, France has a complicated local government system. The major levels of the French system are:

- approximately 36 000 communes;
- approximately 3500 cantons;
- 96 departments; and
- 22 regions.

Additionally there are numerous organizations that group together neighbouring communes in a policy known as *intercommunalité*. These groupings can manage a number of local services such as water distribution, refuse collection whilst others concentrate on the implementation of local structure planning. The decentralization laws of 1982 gave new responsibilities to local authorities and thus new powers that have significantly modified the landscape of French local authorities. GIS technology has erupted on the scene as part of this dynamic change in French local government.

The examination of GIS in towns and Metropolitan Community Councils and in départemental and regional authorities reveals the specific details of the French model. To this end the main features of the relation between local authorities and how they influence GIS implementation will be examined next.

GIS in Towns and Cities

The BDU (Urban Data Bank) experiment

The development of the BDU linked to the data management of urban data has to be put in a broader context of the computerization of towns and cities. The development of processes for the acquisition and the restitution of localized information for municipal authorities was not part of the original developmental core of municipal information technology in France. This core was formed from accountancy needs on the one hand and telecommunications on the other. The first information technology applications were generally linked to the establishment of electoral registers, the local government payroll, etc. Only a few authorities included, from their first information technology master-plans, an application that was described at that time as the automatic management of plans.

During the 1980s there was a manifest and fundamental evolution of priorities and the development of true urban applications. Other issues have been taken on board other than the management of urban services (such as sewerage, public lighting, refuse collection). Dupuy (1992, p. 96) noted three such issues as 'the level of service offered to the citizen, the economy of the urban system and territorial reorganization'. Without doubt it is within this spirit that the new developments of GIS should be seen. However there is always the weight of the past, the old failures and the organizational cultures that are still important in the orientation and the practical day-to-day use of these systems. One cannot talk of the trends in the use of GIS in French towns and cities today without taking into account the traces of the past.

It is generally accepted that the development of GIS in French towns can be divided into three stages: the 1970s, the early 1980s and the late 1980s to the present day as in the earlier discussion.

The 1970s   There are few forerunners to the BDU. Some towns had started to develop computerized systems using localized data from the early 1970s. These experiments have paradoxically served as both the real world test of BDU feasibility (the most recent creations have taken into account the experience gained by these pioneers) and as scapegoats for those who have doubted, sometimes justifiably, the usefulness of the initial results.

Paris had the first French BDU. Created in 1969 at the initiation of APUR (the Paris Planning Workshop) and IAURIF (the Institute of Planning for the Ile-de-France) it contained information on the built environment, land parcel data and socio-economic data from INSEE. In this period there were four other experiments of note:

- The 'Underground Cadastre' of the City of Lille: the reference geographic information was developed at 1:200 scale, based on a survey of the road network and information from the subdivision plans. As a result of choosing this very detailed scale, data capture for the 1800 km of roads took 15 years! The Lille experiment has exemplified the difficulties of the BDU concept and it has experienced many problems because of its pioneering nature. In 1987 the Greater Lille council commenced data capture for land parcel management that is projected to be completed in 1995. This will be the year (23 years after its start) that the reference topographic structure will be in place.

- The Cadastre plan survey and the topographic complement for the City of Marseille: BDU implementation dates from 1972 and it was one of the first. Data capture for the cadastre started in 1973, and a formal agreement with the DGI in 1991 makes it recognized as operating to the accepted standards of the DGI. The first topographic map at the scale of 1:200 scale was produced in colour in 1978 (a considerable feat for the period). The Marseilles experiment has been, without doubt, one of the success stories of broad-based BDU development. It continues to be operational and expanding with new applications.

- The development by the DGI of the first digital cadastres for the cities of Paris, Lyon, St.-Etienne, Bordeaux, Montbéliard: these projects were completed in the mid-1980s and were the testing ground for the launch of the PCI (Digitized Cadastre Plan) for the whole of France in 1991. This project will probably go on for 30 years.

- The implementation of the RGU (the General Urban Survey) in the major cities: other cities and Metropolitan Community Councils are included in this large project (such as Lyon and Toulouse). It is important to understand the context for these experiments. The early years of the decade were marked by numerous projects led, sometimes without consultation and co-operation, by different agents who sought to produce basic geographic reference information. Without going into detail, it might be useful to recall the digital inventory of roads and place-names (FINATO then RIVOLI) in 1971 by the DGI, the General Survey mentioned above (RGU) which started in 1973 under the aegis of several ministries and the INSEE to produce a correspondence between city blocks and the road network and other developments such as the creation of national statistical files (SIRENE).

The main concerns of this period were not subject-oriented but technical in nature. Computerization was perceived as a chance for the automation and the speeding up of boring repetitive work. This is why the initial objectives were couched in terms of reducing time to locate information and to up-date plans. It is also notable that the first BDUs were more often the responsibility of computing departments on the one hand and mapping departments on the other. This somewhat anecdotal evidence explains in part the principal choices that were made some 20 years ago in terms of reference data (such as data type, scales, precision, geographic reference framework) and the types of application chosen.

The lessons learnt from this period are both structural in nature and specific to the period. First the specifically contemporary lessons arose through the major difficulties encountered by the pioneer cities (delays in completion and overspending of budgets) that can only be explained by reference to the technical environment of the period. The absence of high performance tools, suitable data capture equipment and suitably qualified staff played an important role in the derailing of these early experiments. However, they do not in themselves explain all the difficulties. It is necessary to consider underlying structural causes for which the lessons to be learnt are still relevant today. First, there was the general underestimation of the time required to capture data that often did not exist or was incomplete. Additionally, there was a desire to manage the totality of the process in-house without coordination or partnership with external organizations. This was also true for the national projects, such as the RGU and the PCI, that were destined for use by municipal organizations but were pursued without consultation with prospective users. This created a tension between the proposed product and the expectations of clients that sometimes provoked the abandonment of the project.

The early 1980s   This was a period of transition during which some of the BDU pioneers continued to develop and strengthen their databases (such as Paris and Marseille) whilst others evolved towards the constitution of new databases. The reference database was then a 1:200 scale map of the road network (such as for Grenoble and Valence). The emphasis of such a BDU was directed towards the production of basic plans for engineering works and the location of the underground utility networks. The BDU in Toulouse is an example of this emphasis that was started in 1984 and the 1:200 scale map was completed in 1989.

The evolution of the computing hardware allowed the easier modernization of plan production and of CAD (computer aided design) applications. The applications from this type of BDU were essentially biased towards the production of data for the management of technical services and were not intended to be an information system for the authority as a whole. Once again, only technical aspects of urban planning were included and localized information was used to produce plans. In practical terms: 'it seems that interventions within urban areas can be often reduced to management, that is to a calculating process guided by financial constraints. In the same way, the BDU has become a tool for urban management in the strictest sense of the term. The city caught in the municipal BDU, does it not appear as an object to be managed and no longer as a place and an entity full of diverse projects?' (Aillaud, 1992, p. 21). This is a period of transition from which new concerns develop accompanied by noteworthy progress in information technology and notably the arrival of the first GIS in Europe and France.

The late 1980s to the present day   There have been a number of external factors that need to be recalled in order to understand the recent evolution of the use of geographic information in urban authorities. First of all, there is the impact of the changes in the responsibilities of urban authorities as a result of the decentralization laws. This is the impact on the general self-perception of the authority as much as on the make-up or evolution of the BDU. In effect: 'from this time on it is the communes that are responsible for plan-making, development control and development projects, while the State remains guarantor for the overall balance and for projects that are in the general interest. Consequently this has helped to diversify and multiply the number of project clients as much as it has led to a new expression of needs if not of totally new needs' (Wolf, 1993, p. 14).

The same author underlines justifiably the changes in the economic and technical environment that partly explain the change in expectations and outlook. These modifications include: the need for a planning system that is more flexible and more modular in a period of weak economic growth; the taking into account of expectations arising through advanced technology; and the need to manage the environment and the willingness to conserve the existing environment. The fundamental change in the operational world has been the realization by certain decision-makers that 'the priority is no longer to produce precise plans for engineering works but the implementation of management tools for urban data' (Fournillier, 1989, p. 28).

By a happy coincidence (that was not necessarily by chance) the emergence of data management tools with functionalities by far superior to the tools of the previous decades onto the French market happened from 1985. The BDU were redirected in a new sense by trying to combine the functions of the earlier period (needed and useful for management) with the analytical functions that only GIS offered. A good example of this type of evolution is the BDU of the Greater Lyon council that in 1985 started a broad-based feasibility study to initiate a 'new formula' BDU in 1987. The choice of GIS software (APIC) demonstrates the desire to develop a multi-user system that is a clear evolution from the purely management-based BDU.

The Lyon example shows how the two aspects (management and analysis) can co-exist within the same project. A detailed analysis of the development of this project shows two phases. The first phase is the constitution of a basic database (notably the Cadastre) and the development of management applications (such as public lighting, master-planning and roadside trees). This is followed by a second phase (laid down in the information technology master-plan of 1992) that includes a number of planning-related topics: property and land ownership observatories, urban ecology observatories and management of enterprise zones.

The change is significant and carries with it new possibilities for how geographic information is used in operational analysis. These new possibilities that are in part related to the new software approaches linked to GIS, brought with them a wave of BDU projects between 1989 and 1991. An in-depth study of these new projects is not currently possible due to their newness. However, a number of clear trends are evident and many of these are compatible with the evolution of the combined use of data management and strategic decision-making.

An example of such a multi-use BDU is that of the Greater Brest council. At Brest it was the arrival of a municipal team in 1989 with economic and planning objectives for the city (including a strong policy emphasis towards the environment) that gave the BDU a predisposition towards analysis. The base data was a collection of information from the Cadastre and topographic sources at 1:5000 scale. The chosen software was a market-leader GIS software (Arc/Info) and this selection broke the tradition of internally developed hybrid products that were often unsuited to the needs of the authority. This choice was also linked to the fact that unlike previous experiments in this field, precise localization was not the main priority of this system. The first results showed that the system was subject-oriented with maps of census data at the level of city blocks, urban heritage and landscape plans. The use of these systems to locate businesses on the basis of postal address within the municipal area is a good example of a new focus and of emerging applications for which the BDU is considered as a means of assisting decision-making with the help of GIS.

In the space of a decade, the introduction of GIS into the urban arena has profoundly changed the vocation of the BDUs. The major implications of this have been mainly in the evolution of the scale of basic reference. The basic plan of the road network is no longer an obligatory development stage and the adopted scales are more often compatible with the work of urban developers and planners. This has had a significant bearing on the time required and the cost of data capture. The long lead-time required for data capture 20 years ago will become part of an unfortunate historic trial run.

The multiplication of possible applications forces the BDU data to be put at the disposal of a number of different departments and partners. Initially embedded within the computing department and then in the topographic and computing department, the BDU today is shared among a number of other users. This is a significant change from recent years, leading to the systematic gathering of partners within the BDU project (such as infrastructure managers, the IGN and the DGI) who collectively form a network of agents linked together by their participation in the project.

The BDU are no longer only found in the most important cities (even if the trend is for the level of equipment to follow the size of the town, see below). It is now clear to a number of specialists that 'the potential for the decade to come is linked to the acquisition of these systems by small and medium-sized towns in large numbers at a price that they can afford [and thus to pass from] experimentation and innovation to the industrial style management of geographic information' (Daull, 1992, p. 3).

Current trends in GIS evolution for urban local authorities

GIS evolution in urban local authorities is, as yet, difficult to evaluate with precision. A number of studies regularly seek to describe the state of diffusion of this market sector which is in a phase of rapid development. The results are by necessity only partially due to this dynamic change but many cities are currently in a state of transformation from automated drafting systems (such as Autocad) to GIS software. Hybrid solutions still exist, such as GIS that is installed in either only one or few departments, and there is still little evidence that these systems are used effectively. At this point only a number of trends can be indicated. These trends have been selected from elements of a survey carried out by EuroVista (1992). These elements have been blended with other information provided by the Association for IT in local government (ATOLL) and the Géomatique Observatoire of the consultants IETI.

The list of authorities in Table 10.1 is only indicative but underlines the recent nature of GIS applications in French local government. It must be emphasized that the table is only concerned with the adoption of GIS as in the majority of these municipalities information technology tools already existed. Many of the pre-existing systems are still operational. In some cases, data have been transferred from the old system to the GIS (for example, Nice where data was transferred from GPG to Geodis) while in others there is a co-habitation of the two systems (for example Rennes with both Ascodes and Arc/Info).

The second trend is the obvious relation between municipality size (thus potential resources for investment) and the acquisition of a GIS. This tendency should stabilize in the years to come with the arrival of GIS on relatively cheap personal computers that are powerful enough to respond to the needs of urban subject-oriented applications. This trend is best illustrated by the EuroVista figures.

The current situation is the result of the recent transition period. The relation between the municipality size and the level of GIS investment is striking. Nearly all

Table 10.1 Date of acquisition of GIS for selected urban authorities
in France

Municipality	Population	GIS acquisition
Toulon	172 000	1985
CU^a Lyon	1 200 000	1987
Dijon (District)	230 000	1988
Mulhouse	110 000	1989
Nancy	100 000	1989
Metz	120 000	1989
Albi	50 000	1990
CU Mans	190 000	1990
CU Brest	220 000	1991
Angers	146 000	1991
Perpignan	110 000	1991
Nice	350 000	1991
Strasbourg	425 000	1991
Reims	210 000	1992
Castres	46 300	1992
Caen	110 000	1992
Rennes	203 000	1992
CU Dunkirk	215 000	1992
Roubaix	99 000	Under development
Orleans	108 000	Under development
Amiens	136 000	Under development

^{a} Communauté Urbaine.
Source: M-L. Coudun (1993) and RDI (1992).

of the large municipalities with more than 100 000 inhabitants are either already equipped or have a project in hand. The difference in comparison with smaller municipalities is very clear. It is important to note the number of municipalities who are currently considering the acquisition of a GIS. It is obvious that these figures will change considerably as medium-sized municipalities acquire the capital equipment. It is interesting to note the principal problems cited by the project managers (regardless of municipality size) are in order of priority financial and organizational; technical problems are only cited last and this can be taken as proof of the rapid technical progress in recent years.

Small towns are not featured in Table 10.2, although there are a number of GIS projects being undertaken in small towns, often with local partners (such as local surveyors). Because of this, it is extremely difficult to have an overall view of diffusion because such projects are essentially single subject-oriented applications on personal computers within the organization. It is clear however that the very large number of small towns in France make this section of the market extremely important in terms of GIS diffusion. Within these small territorial areas where there are fewer organizational problems than in large authorities, it is possible to develop operational PC-based GIS linked to statistical packages for which the scale of the data is compatible with decision-making for planning.

Table 10.2 Plans for GIS acquisition by municipality population size

Municipality size (no. inhabitants)	No plans for GIS (%)	Considering GIS (%)	GIS project planned (%)	Equipped already with (%)
>100 000	0.0	8.8	23.5	67.6
70 000–100 000	25.0	12.5	31.2	31.2
35 000–70 000	13.9	41.6	18.0	26.4
20 000–35 000	25.4	42.4	23.7	8.4

Source: EuroVista (1992).

The evidence available suggests that the general trend is for an increasing GIS diffusion within urban local authorities. Although in some cases this diffusion represents an evolution of existing BDUs, in the majority of instances it includes entirely new developments. Even the concept of the urban database is under transformation as it is enriched by the development of clearly defined and subject-oriented applications that respond to a wide range of users. There is a conjunction of a demand for increasingly diverse uses that blend management and planning decision-making with the maturity of GIS in how it structures and processes geographic information. This conjunction allows the development of a BDU/GIS that places a greater emphasis on the analysis of the urban environment in its broadest sense.

GIS and knowledge of the urban environment

Do the new expectations seen today in the opening up of the BDU to applications that are more subject-oriented than before allow us to foresee an improvement in the operational use of information from purely data processing? Are the data types different from those that were used previously and if so can these previous data types be used for taking strategic decisions? Wolf (1993) justifiably underlines that data collected today by municipalities can be classified in three ways: 'natural' data such as site data, sub-soil conditions and landscape constraints (these data increasingly need to be taken into account for ever increasing statutory environmental concerns); 'artificial' data that will include all infrastructure built within the municipality both above and below ground; and 'abstract' data which include the socio-economic indicators for the municipality inhabitants. With the increasing reliance on the logic of the information system these data types are gathered and stored in the same environment. However a satisfactory technical performance is not a guarantee of the optimal use and the municipality often finds itself confronted with the same methodological problems encountered by others when combining the analyses of 'natural' and 'abstract' data.

The municipality is in a particular position as it is responsible for both technical management needs and strategic decision-making for land-use and urban economic planning. The fundamental problem that confronts the municipality is tied to the many different levels of decision-taking and the impact that this has on geographic information in terms of what is captured and what is put into the reference base layer.

A problem that seems to exist today is to identify what is strategic information for the city? How can the decision-making process be introduced into the digital database? Who is to structure strategic information for use within a GIS? The current situation is in a state of flux. The realization of the potential of the BDU/GIS within the decision-making process for planning is a recent event. The examples of recent developments show two trends: either to develop first a technical database necessary for the sound management of the urban space, extending this database in a second phase to applications that can support decision-makers (e.g. the BDU of the Urban Community of Lyon); or to develop directly (principally in medium-sized towns) a gathering of information specifically aimed at urban developers (e.g. CU of Brest).

It is also appropriate to ask questions related to the integration of the BDU in surrounding areas. For urban management problems the choice of reference spatial coverage is not a crucial issue if it is not a question of up-dating the database as required with the extension of infrastructure. Within a planning perspective and taking into account the imperatives of multi-scale integration how does one use a BDU that either already exists or is under development? What linkage can there be or should there be between surrounding areas? How does one manage the existence of high information density environments and the rupture that is induced at the territorial limits of the database?

These questions highlight the problem of space that has an unequal information coverage. Before answering this question it is appropriate to underline the apparent incompatibility of a spatially constrained database fixed by the administrative boundaries of the municipality with the larger spatial sphere of influence to be considered for its development policy. Thus 'the relevant framework for a policy of intervention has evolved into that of a wider urban/rural entity, while in the past the relationship between these two types of space has been treated separately within land-use planning' (Asher *et al.*, 1993, p. 143). The potential, from a land-use planning perspective, is to develop a seamless geographic database that complements different administrative and geographic areas and is coherent in terms of strategic action.

From the point of view of the municipality it is the link with GIS developed by higher spatial levels of administration (the département or the region) that holds the key to this necessary complementary characteristic. However, as outlined below the examples of GIS developed at these higher spatial levels are rarely designed in partnership with lower-tier authorities that already have computerized geographic information capacity. They are more often driven by political calculation and designed in ignorance of the existing potential for integration.

One example that does demonstrate good practice is that of the Ile-de-France region. The GIS of the IAURIF (the Planning Institute for the Ile-de-France) for which Paris is the major party covers a large part of the region and makes it possible to satisfy both the requirements of BDU and of a regional GIS. This experience is an example of fruitful complementarity being based on the evolution of the 1969 BDU towards GIS for decision-support. The subject-oriented evolution of the urban database of Paris and its region is one of the better examples of this trend to integrate the GIS tool to meet the demands of decision-making rather than those of management. It was the development of the MOS (land-use inventory) in 1980 with subsequent up-dating that has given the Parisian BDU a planning dimension. In 1990 the GIS dimension was emphasized with the development of the RGIS (Regional GIS) for the Ile-de-France. The objectives of the RGIS are to allow a reference base map of the communes to be used for planning studies. These studies can utilize different subject-oriented information layers that are available within the database (notably demographic layers), create overlays with the MOS and produce new information such as maps of land-use density. This cross referencing of information is fundamental to this experiment and offers a good example of the potential use for GIS operators (the GIS used in this example is Arc/Info).

The planning dimension of the RGIS is clear with the objective of being the basic reference for property and environmental observatories for example. One of the best proofs of the integrated management/planning use lies in the potential for access to documents from a common database but with varying levels of precision (for example the MOS with either 130, 45, 21 or 11 nomenclature classes) and for planning documents that are simplified for decision-makers.

The interesting features of this project lie in its taking into consideration a regional geographic space in which the Parisian BDU is integrated, and in its double logic of a reference database as well as the basis for the development of applications geared towards planning. In fact the development of the RGIS was carried out in parallel with the existing BDU that continued to operate in gathering and validating general data.

This example shows the extent of the overall considerations that have to be made in developing planning-oriented GIS in which different levels of scale complement each other. This issue is all the more important in recent years with the development of départemental and regional GIS for which the objectives are clearly different from those of the classic BDU and which use different geographic data sets.

GIS at the Higher Levels of Local Government

A review of the experiences of the départemental and regional GIS development is necessarily briefer than that for the BDU. These developments are often more recent. The oldest examples (Hérault and Vaucluse) date from the period 1988–90. But even so the number of systems is still low especially at the regional level. However certain trends are already distinguishable and it is possible to discern from the objectives of the projects the design issues linked to the scale of work and the mission statements of the respective government levels.

GIS in the Départements

The first départemental GIS projects were contemporary with the reorientation of the BDU (i.e. the late 1980s and early 1990s). Currently, the available figures show that an expansion of interest is under way. Table 10.3 by EuroVista identifies the level of GIS diffusion among the 96 départemental councils (conseils généraux) in 1992.

More recent figures (IETI, 1993) show 27 départamental councils equipped with GIS which is 28% of the total. However, the purchase of software is not synonymous with the concerted effort for the development of an information system for the authority. Moreover, a more detailed analysis of the list reveals that certain départements are only equipped with software based on personal computers. This demonstrates that the GIS is aimed at a specific application possibly within a solitary département and it is not designed to be part of an inter-service infrastructure. From the evidence only a few départments have taken the step to develop a broad-based system (Hérault, Vaucluse, Savoie, Var, Maine-et-Loire).

Table 10.3 Plans for GIS acquisition in the départements

Plans for GIS	Percentage
No plans for GIS	31
Considering GIS	39
Planning a GIS	11
Equipped with a GIS	19

In general terms, the GIS being developed at the départemental level are in support of the statutory responsibilities of these authorities, which include the management of roads and public works, school transport, and property management and disaster management (fire and pollution). Often, the primary objective of départemental GIS is to act as a repository for information that can serve the needs for internal studies and also the needs of municipalities notably with regard to the establishment of a POS (master-plan), road works, and networks in general. Thus with a logic that recalls that of the BDU, the capture of the Cadastre is currently the most widespread application. Such applications are the subject of a formal agreement between the DGI on one hand and the communes on the other. The départemental council plays the role of counsellor and programme co-ordinator as was the case in the Département of Hérault. In the framework of such broad-based schemes partners often collaborate with external service providers (such as EDF/GDF, France Télécom).

Beyond the management objectives and the difference with the first generation BDU the need for instruments for handling geographic information linked to the economy or to statistics and thus to the evaluation and management of actions (notably subsidies) is made clear by the départements even if this is currently only expressed in terms of needs rather than existing applications. As with everything else this is currently in a state of transformation. The recent creation of a number of economic and demographic observatories within départemental councils will surely lead in the years to come to an evolution similar to that of the BDU where the largest part of the geographic information will be focused for strategic decision support.

Even if the technical management component is largely represented in the existing départemental GIS projects the features of the geographic information that is used with the exception of the Cadastre is noticeably different from that of municipalities. Logically the priority data held by départemental GIS is at large and medium scale (1:2000–1:25 000). Larger scale coverage is provided by the Cadastre but the remaining data come more often from the digital databases of the IGN.

The major needs of départements in terms of digital data are of two types. The first type is data for technical management: this is primarily the Cadastre data; the road network data (these data are used for the optimization of school transport); and particular network data such as fire prevention routes. The second type of data relates to more analytical applications and is essentially environmental data (for water resource management, pollution control and geology). It is interesting to note that satellite data are of particular relevance at the départemental scale in vegetation cover maps (as in the Département of Var) and in the management of fire risk for example.

Some of the départemental GIS currently being installed are experiencing a number of difficulties that recall those of the BDU 20 years ago. Unfortunate experiences like that of the Département of Vaucluse are already part of the history of départemental GIS. However, the essential problem comes less from an incapacity of the GIS tool to meet the demands than from a culture not yet used to handling geographic information, and the absence of discernment between technical management objectives and those of the implementation of a system compatible with decision support in planning.

Currently decision-support applications are little more than desires rather than operational concrete programs. This is regrettable since the character of these authorities would benefit considerably from using localized information analysis as a support for spatial decisions (for example rural infrastructure programs, the definition of environmentally sensitive areas, the construction of schools and the drafting of départemental transport policies). The logic for geographic information use is strong in départements that are highly urbanized or in départements where there is a strong risk of natural disaster (fire, risks linked to mountainous environment, pollution and in coastal départements). Surveys show a cruel absence of GIS projects in areas with the least economic growth for example. The pattern of evolution which is outlined is like that of the BDU with the progressive establishment of a dichotomy between authorities that have the information tools for coherent management and planning for whom the digital data are captured and available and those who will remain at the margins of the diffusion of this advanced technology. This problem is not the same for the regions for which GIS is often perceived as a tool for planning and additionally as a tool to promote the image of the authority.

GIS in the Regions

Even though regional authorities have only recently become involved with GIS, they make an interesting area of study. In 1993, 10 of the 22 regions in mainland France had been equipped with GIS software [Franche-Comté, Bourgogne, Centre, Nord Pas-de-Calais, Languedoc-Rousillon, Provence Alpes, Cote d'Azur, Ile-de-France, Midi-Pyrénées, Haute Normandie: IETI (1993)]. Of these 10, only a few are at the model stage (notably the SIGMIP in the Midi-Pyrénées region) with the remainder either being at the stage of the technical or financial feasibility study.

French regions have generally limited financial resources, and their institutional role is principally one of synthesis, coordination, and programming at the regional level. Hence, their GIS projects are not as broad-based as those of large cities and départements. On the other hand, the planning role of the region (regional structure plans for transport and for tourism development, etc.) predisposes them to the implementation of subject-oriented GIS where the technical management aspect features as an auxiliary function.

It is currently still not possible to discern the great trends of regional GIS due to a lack of operational systems. The opportunity exists to implement tools at this strategic level that could be excellent examples of the use of digital databases for planning. The scales of interest to the regions (1:25 000 to 1:100 000) lend themselves to the establishment of decision-support documents. It would be a shame if the current trend, that of the implementation of regional data servers, would be the unique aim of regional GIS. Although this would be an attractive option, given the degree of partnership that it would induce amongst the various agents (state departments, research establishments and other local authorities), the ultimate aim must be the appropriation of the system within the regional council, through the various regional services up to the decision-making level.

The regional level has recourse to data at medium scales (the BDCarto of the IGN, the National Forestry Inventory). It could equally be included amongst the users of data produced at higher spatial levels such as the database produced by the CORINE program at the European level. The possibilities are interesting especially when one considers the national or European levels. The integration of regional GIS in a concerted national or supranational system is an idea whose time has come. The problems of integration have justifiably been underlined by stating that 'the lack of coherence between [national or regional GIS] drives one to imagine supranational GIS susceptible to allow international comparisons or at least transnational studies made necessary by the internationalization of economies and policy' and one can add 'the result is not always convincing as the increase in geographic scale tends to lose the finesse of the indicators under consideration' (de Gaudemar, 1992, p. 1031). The constitution of a European GIS for the environment (such as CORINE) or as a reference database (such as the EUROSTAT GISCO program for example) needs to be taken into account in setting up regional GIS.

This European dimension, specific to the regional GIS, has started to be taken into account at the national level in France. At the end of 1993 a seminar was held in Toulouse entitled 'GIS and European regions' that gathered together decision-makers and specialists as well as representatives of the European Union. The conclusions of this seminar put the accent on the obstacles linked to the supply of geographic data as much as on the need for a better compatibility of data between systems. The cost of digital data over an extended area remains an important obstacle and the users of regional GIS emphasized the need for an improved flexibility of databases for spatial applications that are specific to particular areas. The need for inter-regional cooperation was clearly signposted even if the solutions to the implementation of such cooperation have yet to become operational.

The regional level presents the potential for intervention at scales of action more directly compatible with the strategic decision-making level of local authorities. The necessary information is less a support for technical operations than a support for decision-making for regional planning. Applications can be developed with partners involved in planning in various fields. Thus, for example, the first applications chosen for the SIGMIP in the Midi-Pyrénées Region includes an application for the 'industrial environment' and another on 'town planning and infrastructure'. These two applications were developed in collaboration with a variety of regional organizations and are the proof of an objective to provide an instrument for the regional policy implementation especially linked to planning.

In this spirit the link between the implementation of GIS and the execution of the 9th National Plan in the regions is clearly noted as in the case of Midi-Pyrénées, where 'it appears that from now on, the execution of the 9th National Plan in the Midi-Pyrénées, will fully draw advantage from the existence of a geographic database covering the whole of this region, notably for priority action identified by the Regional Planning Conference: the development of a regional policy for land use planning, the definition of a regional structure plan and an urban framework plan; an acceleration of policies to open up isolated communities through the development of transport networks; the development of a quality agriculture and a balanced development of rural areas; the management of the environment' (Scot Conseil, 1993, p. 8).

The success of regional GIS in France will result from a capacity to distinguish themselves from the BDU and the management applications of départements in order to consider geographic information in a strategic perspective and its applications to regional planning. It is necessary that elected members are made aware of the potential of GIS in order to assure the implementation of regional GIS that has the potential of being a pivot between local and European levels of government.

CONCLUSION

In the limited discussion concerning the diffusion of GIS in French local authorities it is important to remember that the success of this diffusion is partly linked to the development of projects that join together different factors in collaboration with local government. The particular details of the French situation (the clear division between the powers of the State on one hand and the various levels of local authority on the other) and the re-equilibrium of powers that is currently taking place make predictions difficult.

This chapter has given an overview of the extent and rapid diffusion of GIS in local government in France. However, it has also signalled that the implications of these developments for planning and territorial management have yet to be fully appreciated both at national and local level. There is a fundamental need for a general reflection on the role of geographic information beyond the simple figures of the potential market for GIS. The hard implications of the implementation of information systems in local government has to be part of a global policy framework where specific interests are not in conflict with the general interest for a harmonious development of the national territory. The full dimension of geographic information as a support to local planning action has to be understood by local government at the decision-making level. It is necessary to promote a geographic information culture that includes more than just the technical aspects of its manipulation.

In France as elsewhere, urban space is the privileged domain of GIS use. The long tradition of urban data bank implementation contributes to the increasing diffusion of this technology identified in this chapter. At the urban level, there is a concentration of localized data appropriate to the use of GIS, while the high profile technical management component of the work of city councils has allowed the early imple-mentation of GIS. This has subsequently provoked specialist developments that have sometimes been difficult to transform to the needs of applications directed at planning. In France the diffusion of GIS that is generally urban in character will no doubt soon be complemented by the implementation of GIS in the départements and the regions as well as programs concerning inter-communal structures. The rapidity with which these developments are taking place, requires continued efforts for their monitoring and evaluation.

REFERENCES

AILLAUD, V., 1992. Que sont devenues les banques de données urbaines?, in *SIGAS*, **2**(1), 15–27.

ASHER, F. *et al.*, 1993. Les Territoires du Futur, DATAR/Editions de l'Aube.

BOURSIER, P., 1993. Problèmes techniques liés aux données dans la mise en place des SIG, in *Actes du Séminaire ENGREF: le Développement des SIG*, doit-on se Préoccuper des *Données?*, Montpellier 10–11 Mars.

COUDUN, M-L., 1993. La France des SIG, *Genie Urbain Aménagement et Territoire*, No.397, mai 1993, 10–12.

DAULL, B., 1992. Éditorial de *Génie Urbain*, Mai, No. 397, 3.

DE GAUDEMAR, J.P., 1992. L'aménagement du territoire, in Bailly, A. S., Ferras, M. and Pumain, D. (Eds) *Encyclopédie da la Géographie*, pp. 1023–43, Paris: Editions Economica.

DENEGRE, J., 1989. Les enseignements tirés des travaux de la commission topo-foncière du CNIG, in *Les SIG: Actes du Séminaire CNIG-AFI3G*, 22–3 Novembre, Paris.

DIDIER, M., 1990. *Utilité et Valeur de l'Information Géographique*, Paris: Editions Economica.

DIDIER, M. and BOUVEYRON, C., 1993. *Guide économique et méthodologique des SIG*, Paris: Collection Géomatique Editions Hermès.

DUMONT, J.M., 1992. Les données du Cadastre, in *Actes de la Conférence MARI*, Paris, 22–4 Septembre, pp. 149–53.

DUPUY, G., 1992. *L'informatisation des Villes*, Que sais-je No. 2701, Paris: Presses Universitaires de France.

EuroVISTA, 1992. *Les SIG dans les Collectivités Territoriales*, Paris: Nouvelles Editions Européennes.

FOURNILLIER, J.M., 1989. Problématique des villes françaises en matière de banques de données urbaines, in *Actes du Séminaire CNIG/AFI3G*, 22–3 Décembre, Paris.

IETI, 1993. Personal communication, *Observatoire Géomatique*, Mâcon.

JOURNAL OFFICIEL, 1985. Décret No. 85-790 du 26 Juillet 1985 relatif au rôle et à la composition du CNIG, Paris.

LAMY, S., 1992. L'information géographique numérique â l'IGN, in *Actes de la Conférence MARI*, Paris, 22–4 Septembre, pp. 141–8.

RDI, 1992. *Dissémination des SIG ainsi que leurs Applications dans les Collectivtés Moyenne en Europe et dans le Monde*, rapport intérmédiaire, Programme SPRINT, CEE.

SCOT CONSEIL, 1993. Dossier Technique: SIG régionaux et départementaux en France, in *SIG et Télédétection*, Mai, **7**, 7–9.

WOLF, M., 1993. Outils et méthodes pour connaitre, concevoir et gérer la ville, *Genie Urbain Aménagement et Territoire*, No. 397, mai, 1993, 14–18.

CHAPTER 11

Poland: methodological doubts and practical solutions

MALGORZATA BARTNICKA, SLAWOMIR P. BARTNICKI
and MAREK KUPISZEWSKI

INTRODUCTION

This chapter discusses the development of Geographic Information Systems (GIS) applications in local governments in Poland over the last 5 years. There are two basic methodological approaches in the research into the distribution of information technology (IT) and GIS in local authorities. The first one is to conduct questionnaire research across all or a selected category of IT/GIS users. This was pioneered in the classical American study of the use of computers in local authorities (Danziger et al., 1982) and in recent research in the UK (Campbell and Masser, 1992). The questionnaire approach usually gives a wealth of information and a good basis for the general, country-specific characteristic of the process of diffusion of a new technology. As noted in the introduction to this book the differences in the number of local authorities as well as country specificity may make such an approach difficult if not impossible to implement in all cases. The second approach focuses on the investigation of selected case studies and is a lot more selective. It gives a less structured and more casual knowledge and usually does not allow for generalization.

In the case of Poland the large number of local authorities (in 1993 there were 2465 local authorities on the lowest level of hierarchy), poor telecommunication and an obsession with secrecy excluded the possibility of a comprehensive questionnaire research. Therefore, this chapter discusses the two competing models of GIS dissemination in Poland and illustrates them with some examples.

ADMINISTRATIVE CONTEXT

The introduction of GIS into central and local governments depends on a number of factors: who owns the information (Cassettari, 1993), what is the management's awareness (Mahoney, 1991) and quality of the leadership (Masser and Campbell, 1991), what are the organizational structures (Masser, 1993) and consequently the

Table 11.1   Characteristic of the administrative division of Poland at 1.1.1993

Level of administration	Number of units	Population of largest unit	Population of smallest unit	Average size of unit
Regional	49	4 013 211	248 486	784 043
Subregional	327	n/a	n/a	117 487
Local	2465	840 088	1158	15 585

Source: Central Statistical Office, 1993.

structure of the administration. In the case of dissemination of GIS in Poland the latter factor has a decisive impact on the whole process. There are currently four levels of administration in Poland (see Table 11.1):

1 Central, where only central (state) governmental administration exists;

2 Regional (województwa), where both representation of central (state) administration as well as in some cases (i.e. capital city of Warsaw) self-governing administration coexist;

3 Subregional–administrative areas (rejony) which are located between regional and local administration and which carry out some tasks of central (state) administration delegated down the hierarchy and finally;

4 Local, where only self governing administration exists.

The distinction is being made here between governmental administration responsible to Parliament and self-governing administration which is elected locally and is to some extent independent from central government.

   The tasks of various levels of administration are at this stage very difficult to specify as the whole system is being modified. The government of Prime Minister Suchocka (1992–3) started the process of decentralizing power by devolving various responsibilities down the hierarchy of local administration. As a result, some local governments have taken certain tasks which in the past were in the hands of upper level administration, while some have chosen not to do so. For example around a quarter of all schools are supervised by local level of administration with the rest being supervised by regional governmental administration. The process of decentralization and creation of subregional administration was rapidly stopped by the postcommunist government of Prime Minister Pawlak. In effect we observe slow re-centralization and strong pressures on local governments to defer to regional governments.

   The process of change, and the existing administrative duality (state versus self-government administration) cause some confusion and friction in practice and are exacerbated by the different objectives pursued by central and local administrations. It is the contention of this chapter that the development of GIS in regional and local authorities has been driven by this duality. In fact two general models of dissemination of GIS can be formulated. For the purpose of this debate we will name the first model 'top-down' or 'central' and the second the 'local' model.

THE CENTRAL MODEL

The Central model was established in 1991 by the Main Surveyor in the Ministry of Spatial Economy and Construction and is documented in Piotrowski (1991) and Ministerstwo Gospodarki Przestrzennej i Budownictwa (1993a,b). Its main aim was to define the legal foundations and organizational structure for GIS. One of the most important statements in the document is that 'GIS constitutes an element of organizational infrastructure of the State' (Piotrowski, 1991, p. 4). This assumes on the one hand, a leading role of the State in the creation of GIS, and on the other hand a key role of GIS in the proper functioning of the State. The document describes the relation between organizations assigned to create GIS and possible users of GIS as a 'server–client' relation to use a computer-based terminology.

Implication of Data Ownership on the Organizational Structure of the System

In the specific Polish situation one of the major sources of the Central approach is the fact that governmental agencies own spatial data. For half a century the institutions which were (and still are) responsible for collection of spatial data were the Departments of Surveying and Geodesy or their equivalents. According to the law they own all collected data (e.g. cadastral information) printed on paper maps 1:500; 1:1000; 1:2000, etc. Other departments located at different levels of State administration, for example, Departments of Planning cannot use their spatial databases for official purposes without the authorization of Departments of Surveying and Geodesy. In such a situation it is obvious that the central model assumes that the creation of any GIS should be based in the Departments of Surveying and Geodesy.

These Departments are supposed to form 'GIS base units' located usually on subregional ('rejon') and regional ('województwo') levels. 'Base units' are responsible for collecting spatial data about parcels, buildings, and infrastructure, at the largest scale (1:500, 1:1000). Various institutions (governmental, self-governmental, co-operative, private) can provide 'GIS base units' with requested information related to the basic maps and for some of them it will be obligatory. Those institutions can form 'associated GIS units', handling specific spatial databases being in their scope of interest (i.e. branch, economic, social, environmental etc.), but also passing them to 'base units' as a part of a general State information system. The rules of how spatial information can be accessed by users other than 'GIS units', i.e. 'clients' of the system, are not clear as yet, but it is assumed that they will have to pay for information, at least partly.

At the subregional level it is assumed that cadastral data as well as various information currently stored in 1:500 and 1:1000 paper maps will form the backbone of the system. At the regional and central level data will be much more generalized.

Standardization

As the paper-based information was strictly standardized ('K-1 instruction'), it seems obvious that digital maps based on such paper maps should also be 'standardized', so information can be passed back and forth through all administrative levels without problems.

Standardization includes various issues: adopting a set of rules to be used in the process of the creation of the system, imposing strict routines on how the system is being created and finally imposing decisions on the hardware and software used by various 'base units'. The scope of standardization was specified by Piotrowski (1991, p. 18) in the following manner: There is a need for:

> ... common legal base, forcing obligatory data supply and price of data, common global coordinate system, standards applied in the creation of databases, unified principles of information input ..., unified regulations for units operation, integration of base GIS units with Departments of Surveying, Geodesy and Cartography, dependent on the State Geodetic Survey

The scope of standardization proposed above is sensible and in theory does not exceed the necessary minimum. It will be pointed out later that if the implementation of the program is successful it may go much further: the software will have to be standardized. Arguments quoted in favour of the central approach are obvious: graphical and attribute data should be in the same format at all levels and training of the operators and users of the system should be cheap and effective. In practice this means that at the central level it would be decided what software to choose (or at least prefer) and subsequently the chosen application would become a 'standard'. At the moment it seems that Arc/Info with Oracle is favoured by the central administration. Even if this chosen software does not fit the needs of final users at local levels, there is evidence to suggest that if central administration invests in the writing of applications and shells for a specific software and gets a good deal on bulk purchase it will create irresistible financial pressure on subordinate units. Wegener and Junius (1991) reported a similar process in Germany.

Pilot Projects

It is obvious that such a gigantic project cannot be implemented instantly for financial, organizational technical and human capital reasons. In order to gain some experience and train staff it was decided to implement some pilot projects.

Lódz implementation

The model implementation of these officially formulated principles was started in the voievodship of Lódz in central Poland (for location of places mentioned in this chapter see Figure 11.1). This is one of the highly urbanized and highly industrialized regions of Poland. The priority task in this exercise was to get a Land Information System covering the whole region and meeting the principles specified above. The region has been previously mapped in the scales 1:500, 1:1000 and 1:2000 on paper, but some of these maps are obsolete. Some attribute data have already been transformed into computerized form with the help of the EWGRUN package (Góraj and Zaremba, 1993) and is usually kept updated. However in order to complete the coverage it was indispensable to survey some parts of the region. Góraj and Zaremba (1993) assess that in the process of the creation of the database it was necessary to update some 15% of attribute data and as much as 40% of cartographic data. These

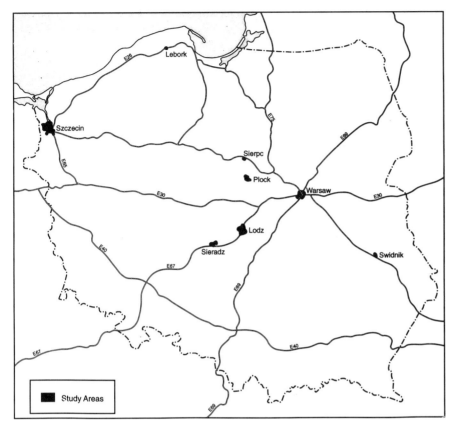

Figure 11.1 Location of case studies

changes resulted in the alteration of up to 70% of information on some parcels! The software selected for the creation of this working version of LIS was AutoCad (Ignaciuk *et al.*, 1993), with the intention to transfer the final product into an Arc/Info environment.

Small scale implementations

Other projects undertaken within the scope of the central approach were those started by the already constituted 'base GIS units' within the Regional Departments of Surveying and Geodesy in Sieradz and Lebork. In both cases the scale of the project was much smaller than in Lódz—covering small towns. The bases for GIS were the digital land inventories existing in either of two systems elaborated in Poland and covering large areas of the country: EWGRUN (in Sieradz) and MSEG (Lebork). Both projects concentrated on downloading the existing spatial data into a system that would enable connection with attribute information. Both these projects were fairly successful in the implementation of the systems.

Preliminary assessment

It is impossible at this stage to assess the results of the pilot projects as no written information is available and those involved in the implementation refuse to give official statements. However we have our reservations about this approach:

- Due to the quality of existing maps and surveys (precision and obsolescence) it would take decades to cover the whole country with digital maps, especially when in many regions paper base maps either do not exist, hold false information, or are very old. Góraj and Zaremba (1993) state that 45% of the area of Poland is covered with good quality cadastral information and that for 14% of the area this information needs to be replaced. For 65% of the area it is impossible to create the geographical part of LIS based on existing information. In many cases the legal position of some parcels is not clear as they were expropriated after the War in breach of the then existing law.

- Apparently both the Polish commercial companies implementing the project and the administration suffer from lack of properly skilled staff and are forced to employ surveyors and computer aided design specialists in order to get the project done.

- It is not clear who would pay for data collection and digitizing. The State budget, in our view, cannot afford this kind of project. Lack of funds slows down the whole process. On the other hand, this sort of project may be financed from local budgets, but the problem here is that local authorities may be reluctant to finance projects in which they will not have their say and which may not meet their needs.

- Not all information covered on base maps is needed by all users, so it is more efficient for final users to collect and update information that is of interest to them (for example utilities companies hold their own inventories).

- The region selected for the pilot project was far too large and complex. The process of learning has been hampered by the difficulties in overcoming the unexpected and numerous technical problems.

- There is a gap between the existing Polish legislative framework and the needs of the project. The pilot implementation apparently suffered from this gap but on the other hand allows for the accumulation of expertise indispensable in the creation of the new law.

The model of the creation of the GIS by the central governmental administration reflects the natural (from the point of view of Central Agencies) tendency to control everything that is happening at lower levels of administration. It stresses the rights of the State, but to a large extent it ignores the needs of many potential GIS users who do not necessarily need the kind of information included in geodetic base maps.

THE LOCAL MODEL

The other model of the dissemination of GIS was called by us a 'local model'. This is because it is fueled with local initiatives developed spontaneously without State intervention. Basically there are three reasons for these developments:

1 The growing awareness of local authorities and growing recognition of the need for GIS development.

2 Local governments, dependent on local electorate, perceive the need to improve widely understood services.

3 A reaction to the lack of decisions at central level.

In this section we will discuss some case studies and point out the advantages and disadvantages of this approach.

Regional (voievodship) Level: the Plock Case Study

The Plock region (voievodship) is located in central Poland about 100 km north of Warsaw. Its capital city, Plock, situated on the east bank of the Vistula River has around 120 000 people and houses the largest oil refinery in Poland as well as a large number of pipelines. The area of the whole voievodship is approximately 5100 km².

The Scope of the System

After several meetings with potential users (staff from Departments of Planning and Environmental Protection) it was decided that digital map at 1:25 000 scale with information typical of standard topographic maps at this scale should be created to meet the identified user needs. Eventually, the information converted into digital form included hydrography, land use, human settlements boundaries, roads, railways, forests, wetland, etc. In all, 69 paper map sheets were digitized and checked during a period of 8 weeks and now the GIS is in constant use in planning departments (road planning and maintenance, utilities analysis and planning, land use planning) and environmental protection (water resources analysis and pollution monitoring).

Practice of GIS implementation: training and institutional change

The staff of the Planning Department was trained by the company which delivered the GIS package. Some problems had arisen already at the preliminary stage, even before the use of GIS really started. These problems were of a basic nature:

1 The fear of the 'unknown' resulted in a negative attitude towards the whole idea.

2 The lack of computer experience even deepened these feelings, as it seemed easier and quicker to do things the usual way, than 'to fight' with computers.

3 The difference between the appearance of paper maps and computer maps was difficult to accept, and was often considered as a proof for the thesis that computer maps are 'all wrong'. For example, the idea of having the names of places in a separate information layer, was hard to accept.

4 The data had to be input manually, and the slowness of the process was discouraging; as the department does not have extra money for data input, it has to be done by its own personnel, that considers itself as too highly qualified for this sort of work.

In fact almost no institutional changes have been made so far as a result of the introduction of GIS in the Planning Department as only a few people were assigned to new duties connected with GIS. Nevertheless, after a few months' period of getting accustomed, the system is now being used. Another problem, of an organizational nature, emerged when the Planning Department tried to get data in digital form from voievodship's Statistical Office in Plock. These data are processed in Warsaw and the voievodship Statistical Offices have data only in disaggregated form, as they do not do any data processing themselves. They can get aggregated data from the Central Statistical Office, but only as a printout or they have to pay for them as does any other organization. An initiative has since arisen to change regulations in this field and enable the use of digital data by different users within the administration of voivodships without charge.

The development of a digital map had an important impact on the data resources in the Planning Department. The rigors of computer work forced a certain ordering of data, clearing of inconsistencies in records, and finding missing information. The main result of the starting phase of a project was an overall improvement in information availability and its proper inventory. This project was meant to be a starting point for creating more detailed computer maps in towns located in the voievodship. The rationale here was to create coherent spatial databases which would ease communication between planners at voievodship level and those at local level.

Software

There was a problem of choosing the best software. The market for GIS software is well developed in Poland so there was no problem of buying a good system but rather to find an appropriate one. Since many Polish local and voivodship planning agencies usually have little or no computer experience, GIS software has to be easy to use [no typing commands, be intuitive (graphic user interface)], be customized to the Polish language, have a powerful relational database with built-in structured query language, be multiplatform and expandable in future, have good communication with other existing software since it has to import and export Dbase, Excel, FoxBase, Lotus, ASCII, DXF, etc., files, be relatively cheap, and have on-line help, manuals, reference books, tutorials, etc., in Polish. The package which perhaps best meets these requirements was MapInfo for Windows which eventually was implemented at the regional level.

Local Level: Sierpc Case Study

When the first stage of creating GIS at voievodship level was finished (i.e. all cartographic data were available in digital form) the next stage of the whole project began in Sierpc, a medium-size town typical for the region. Sierpc is located in the Plock Voievodship, 35 km Northeast from Plock and 122 km north of Warsaw. It occupies a land area of approximately 30 km² and has as approximate population of 20 000 people. The town provides standard municipal services, including maintenance of streets and pavements, water, sewers, storm pipelines, electricity and gas lines. Property ownership is divided among 4000 parcels delineated on the town's base maps, 1:500.

Although it should be recognized that most potential GIS users do not initially know precisely how the system will work and what they will ultimately be able to do with this tool, planners from Sierpc had numerous ideas about its possible uses, and they were really enthusiastic about its implementation. Also they well calculated their financial resources against the costs of digitizing. Several meetings between Sierpc Planning Department staff and IMAGIS, a GIS company, were held to discuss their needs, the availability of data, how the data might be corrected, the possible uses for the system and its future capabilities which could be developed over a longer time. It was decided that during the first stage of implementation parcels and buildings in the town centre (75% of all parcels and 90% of all buildings) should be digitized. During the next stage of implementation the rest of the parcels and buildings will be included as well as utility lines and other information. At the moment more than 70 paper maps at 1:500 are already digitized and a data base concerning parcel and building is being built.

The digital map of Sierpc town centre was developed as a fully functional pilot project to give the Planning Department staff a fully operational graphic and text database with advanced graphic display, SQL possibilities for planning and modelling, with easy access to other databases. The staff of the Planning Department was trained and the expansion of the system, with the inclusion of a new information layer and additional attribute data will follow. Sierpc has taken the opportunity to develop a GIS which is affordable even with today's economic turmoil, and which is easy to use, expandable, and customized to the Polish language. Here the choice was easier to make because MapInfo had been already implemented on a higher level and it was working well.

Other Examples

It appears that the local model approach is gaining much attention from local authorities around Poland. After the success of the Sierpc implementation other projects were started. At the moment this type of GIS building—based on MapInfo for Windows is implemented throughout Poland: in Elblag, Kedzierzyn-Kozle, Lebork, Lubin, Pulawy, Sieradz, Siedice, Szczecin, Swidnik, and tens of other small cities inhabited by 20 000 to 100 000 people. An exception is Szczecin with half a million population.

Szczecin

In Szczecin (a harbour city, situated on both banks of River Odra in north-west Poland) a street map was built at first. Later cadastre information was added. The Szczecin Geographic Information System (SGIS) was built with the help of local GIS and mapping companies. This system has helped to recover much information hidden in forgotten databases. Local government officials claim that during the 1994 fiscal year, the identification of unpaid land taxes (which would not be possible without GIS) will earn the city a significant amount of money, possibly as much as 500 000 USD.

Swidnik

In Swidnik, a 60 000 people city in south-eastern Poland, the town authorities, after a trial implementation of GIS in the central part of the City decided to build a 'total solution' system in which all available information (parcels, buildings, sewage system, electrical lines, water pipelines, cable television network, telecom, greenery, etc., etc.) will be included. It seems that local facility maintenance companies are willing to participate in a project. Inclusion of the information from them assures that the accuracy of data input into the system will be high.

Local Approach: Advantages and Dangers

The bottom-up implementation of GIS at the local level, without any external pressure but also without external support, has both advantages and disadvantages. They are outlined below.

Advantages

Projects at local level:

1 are more realistic in scope as they reflect the needs perceived by practitioners of local planning offices, utility companies, environmental protection etc.;
2 are cheaper and faster to implement because information included into digital map layers are only those which are really needed;
3 are more likely to be successfully implemented in existing conditions (equipment availability, personnel qualifications and attitude);
4 allow for selection of appropriate software;
5 will be implemented only by an enthusiastic team with sound leadership (given the amount of problems to be overcome);
6 allow for tailor-made training.

Dangers

The local approach carries also many dangers, resulting mostly from incompetence:

1 Even basic concepts are often misunderstood, for example differences between raster and vector data (the idea of digital map being a product of scanning is widespread in Poland).
2 The choice of software is often incidental, resulting in implementation problems.
3 The financial risk of wrong decisions is high.
4 This approach is characterized by mostly chaotic implementation of different and often accidentally chosen off-the-shelf GIS software by local governments.
5 Transfer of data between various units and up in the hierarchy may be difficult due to software incompatibility.

CONCLUSIONS

This chapter has discussed two models for introducing GIS in local authorities in Poland. Although their aim is the same: to create a computerized system for the acquisition, processing and management of spatial data; they are fundamentally different. They reflect two contrasting approaches towards basic rules of GIS implementation: how, where, by whom and for what sake GIS in administration should be created.

Both local and central approaches have their advantages and limits. Existing experience proves that the best method of implementation of GIS in administration has to involve closer co-operation between local and regional levels rather than integration of activities on higher levels. This co-operation can hardly be incorporated into the centralized model. However where GIS is being introduced spontaneously it was a lot easier to create teamwork and spirit of co-operation as well as find responsible leaders.

Even if 'standards' of GIS are imposed or 'advised' by central agencies, the real job must be done at the local level. Town managers would rather believe and follow the real life examples which work at local level than listen to theoretical considerations about possibilities of exchanging data between different levels of administration. It is hard for those who pay for their GIS (City Councils, Town Managers) to understand why they should first wait for 'standardization' decisions from the higher levels of administration, and then be obliged to give away data collected and be limited in any way in their use by separate institutions like 'base GIS units'.

It seems that all dangers pointed out in the previous section will be overcome in forthcoming years. First, due to the rapid spread of technology (MapInfo alone was sold in hundreds of copies in the last 18 months) and many educational activities both by universities and private GIS companies, basic concepts and ideas about GIS are quickly popularized. Secondly, rapid software development decreases problems with the exchange of data. Thirdly, the increasing awareness of the costs of wrong decisions lead to a greater use of consulting agencies at the initial stages of the project. This reduces the chances of choosing inappropriate software and makes it possible to estimate the total implementation costs before final decisions are made. It is very likely that in the light of the considerations above the local model will gain in popularity in the coming years.

ACKNOWLEDGEMENTS

The authors are most grateful to Dr Alan Grainger who edited this text and improved their English.

DISCLAIMER

The views and opinions presented in this paper are of the authors and not of their employers.

NOTE

For ease of print all Polish names in this chapter have been rendered to English spelling.

REFERENCES

CAMPBELL, H. and MASSER, I., 1992. GIS in local government: some findings from Great Britain, *International Journal of Geographic Information Systems*, **6**, 529–46.
CASSETTARI, S., 1993. *Introduction to Integrated Geo-information Management*. London: Chapman and Hall.
CENTRAL STATISTICAL OFFICE, 1993. *Powierzchnia i ludnosc oraz bezrobotni w przekroju terytorialnym*, Warsaw: GUS.

DANZIGER, J.N., DUTTON, W.H., KLING, R. and KRAEMER, K.L., 1982. *Computers and Politics: High Technology in American Local Governments*. New York: Columbia University Press.

GÓRAJ, S. and ZAREMBA, S., 1993. *Problemy modernizacji ewidencji gruntów na tle rezultatów eksperymentu SIT w województwie lódzkim*, Lódz: SAP.

IGNACIUK, G., MINICH, P., NIEWIADOMSKI, J. and WLODARCZYK, D., 1993. *Technologia wykonania numerycznej mapy Lodzi dla Systemu Informacji o Terenie*, Lódz: Aplicom.

MAHONEY, R.P., 1991. Making senior management aware of GIS, in *Proceedings of EGIS '91*, Brussels 27 March–1 April, pp. 662–70, Utrecht: EGIS Foundation.

MASSER, I., 1993. Technological and organisational issues in the design of information systems for urban management in developing countries, in Cheema, S. (Ed.) *Urban Management: Policies and Innovations in Developing Countries*, pp. 161–71, New York: Praeger.

MASSER, I. and CAMPBELL, H., 1991. Conditions for the effective utilisation of computers in urban planning in developing countries, *Computers Environment and Urban Systems*, **15**, 55–67.

MINISTERSTWO GOSPODARKI PRZESTRZENNEJ I BUDOWNICTWA, 1993a. *Dokumenty programowe krajowego systemu informacji o terenie*, Warsaw.

MINISTERSTWO GOSPODARKI PRZESTRZENNEJ I BUDOWNICTWA, 1993b. *Problemy organizacyjne państwowej sluzby geodezyjnej i kartograficznej*, Warsaw.

PIOTROWSKI, R., 1991. *System informacji o terenie. Program modernizacji (GIS Modernisation Programme)*, Warsaw: Ministry of Space Economy and Construction.

WEGENER, M. and JUNIUS, H., 1991. 'Universal' GIS versus national land information traditions: Internationalization of software or endogenous developments? *Computers Environment and Urban Systems*, **15**, 219–27.

CHAPTER 12

Netherlands: the diffusion of graphic information technology

AD GRAAFLAND

INTRODUCTION

This chapter presents the findings of two studies on the implementation of GIS in Dutch municipalities, with GIS being defined in broad terms to encompass land information systems (LIS) and systems for automated mapping and facility management (AM/FM). The first research in 428 municipalities (Graafland, 1989) conducted an overall survey of the automation within municipalities. Attention was paid to the relation between administrative data processing and GIS. The second research in 45 municipal departments with GIS within eight Dutch municipalities (Graafland, 1992) aimed at testing and adapting a descriptive model, which describes the development of GIS related to administrative data processing in municipalities. The findings of these two studies explore the organizational and technical elements that influence the development of GIS in local government and describe expectations regarding the development of GIS in future based on a normative phase model.

MUNICIPALITIES IN THE NETHERLANDS

Important for the development of GIS in municipalities is the relation between the municipalities and the other governmental bodies.

The government of the Netherlands has three levels:

1 central government, centred in The Hague;

2 Twelve County Councils also called Provinces, and

3 local government, which includes municipalities and districts in charge of a 'polder-board'. In the Netherlands there are about 650 municipalities and a little bit more than 100 districts in charge of a 'polder-board', with specialized tasks like water control. Because of mergers the number of municipalities and polder-boards is decreasing.

The government of the Netherlands is partly centralized and partly decentralized. This means that central laws are of a higher level than local ones, but municipalities are allowed to organize and manage their own housekeeping. A continuing question is what activities are included under the term housekeeping. Central government and the municipalities are the strongest levels. The position of the provinces in the Netherlands is not always clear although in general terms they implement central government policies, control municipalities and have particular duties related to environmental protection. At this moment there are discussions in the Netherlands to establish a regional level.

The development of information systems in municipalities is in general terms functional to their internal management although they are also required by central government to maintain an automated population register. Important for the diffusion of GIS is a national 'large scale mapping project' called GBKN. This project, managed by the central cadastre, has the aim of creating a nation wide digital topographical map at the scale of 1:1000. After many years of work, only a small part of the Netherlands has been mapped in a digital form. This project more or less failed because of the different interests of the participants involved (like cadastre, utility companies and local government). Increasingly the municipalities have gained facilities and become able to make their own digital topographical maps. Now it is possible for participants at the regional level to define their own mapping project and to come to an agreement with the lower tier authorities.

Although a major part of the municipal information system is a matter of their own housekeeping, there are some initiatives to standardize these information systems. Besides standards developed by suppliers, the association of Dutch municipalities (VNG), is developing standards called GFOs, which mean 'functional designs' for municipalities, to formulate definitions of data and describe data models. VNG wants to help the development of more standardized municipal information systems, with more possibilities for exchanging data and hopes to establish the position of the municipalities as the base of governmental data processing. Until now only a few municipalities have really adopted the GFOs. A major problem for standardization of the municipal information systems is the big difference in the sizes of municipalities and the number and skills of municipal staff as 90% of the municipalities have less than 40 000 inhabitants and 50% even less than 12 000 while the four biggest cities have about 475 000 inhabitants on average.

DIFFUSION OF GIS IN DUTCH MUNICIPALITIES

General

Figure 12.1 shows the situation in 1988 in the Dutch municipalities. Almost all the bigger municipalities had GIS, which means at least one system, while diffusion among smaller authorities was much more limited. The municipalities are classified in four size classes. The figure shows that the bigger municipalities started with implementation during the last decade. Surprisingly there are about the same number of starters each year. So within the same size class there are early pioneers and laggards. Extrapolating this data, it may be safe to assume that all municipalities above 50 000 inhabitants will have GIS facilities in the near future.

Related to the automated administrative and process systems (like specific calculation systems and word processing) GIS is just a relatively small part of the whole municipal automation, as shown in Figure 12.2. As may be expected most

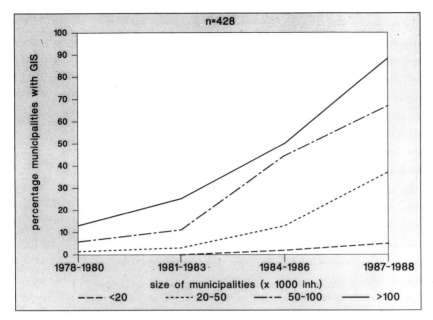

Figure 12.1 The implementation of GIS in Dutch municipalities.

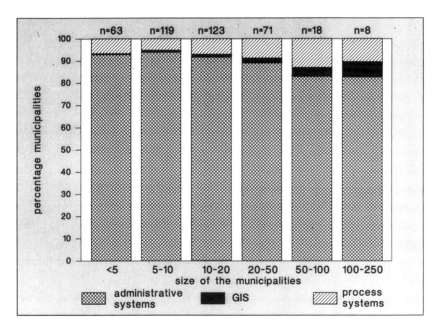

Figure 12.2 Automated applications in Dutch municipalities, 1988

authorities have initiated GIS projects sometime after the establishment of basic computing application. However there are also a few cases of very innovative departments where the start of GIS appears to coincide with or to follow directly after the start of their administrative automation. Research in four medium-sized and

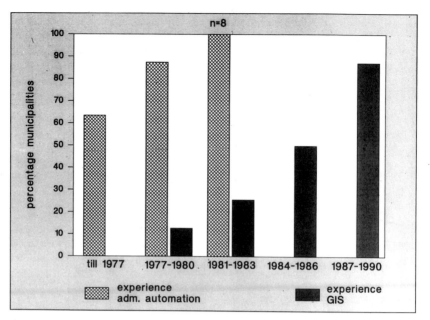

Figure 12.3 Relation between the start of administrative automation and GIS in Dutch municipalities with 100 000–250 000 inhabitants.

four big municipalities gives a more specific overview of the development of GIS (see Figure 12.3).

Organization of GIS

In four of the eight municipalities investigated the co-ordinating departments have a position which enables them to enforce the introduction of a GIS within other departments. The departments hold a high position in the organization's hierarchy. Within other municipalities the departments are autonomous to a considerable extent. As a result of this the influence of the coordinating department is limited.

Regarding the relationship between larger municipalities and computer suppliers there are strong as well as loose ties. Municipalities with a coordinating department with a strong position have a strong tie with a system supplier or software house. They also undergo a fast growth in automation.

Complexity

Figure 12.4 shows the percentage of automation of 91 graphical databases (maps and drawings) found in the municipalities investigated. Graphical databases with simple register functions are relatively more automated than graphical databases with more complex functions (e.g. analyses and planning functions). More than half of the GISs existing in the municipalities investigated thus have a register function only. Only a small part is meant for analysis and a third part for planning. In none of the

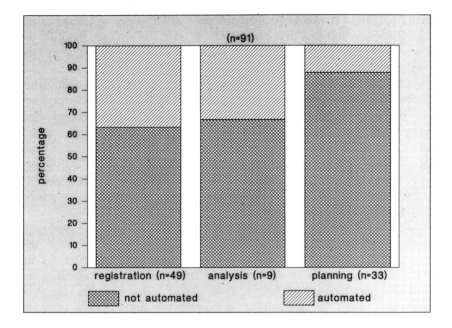

Figure 12.4 Degree of automation of maps, related to the complexity of their function (1991).

municipalities investigated are GIS used at the level of decision support for (strategic) planning. The existing GIS are used for operational activities, including register functions and (structured) calculations, like surface calculation.

Favourable Elements for the Introduction of Graphical Automation

The interviews conducted in the eight case-studies confirmed that in general organizational issues are more critical than technological or data related issues for the successful implementation of GIS. The mentality and the involvement of personnel, management and councillors are especially important. The involvement of staff helps to get the automation accepted in their own department. Departments without experience of GIS, but preparing GIS, appeared to be more concerned with technological problems. But in the case studies involved with increasing the scope and complexity of the GIS, organizational problems were much more prominent.

Development Process

The evidence collected shows that most case-study authorities at the early stages of implementation were concerned with map and drawing production and in particular with increasing the quantity of maps and drawings available in the system. There was a perception that the automation of map production would lead to an increased quality of the maps available (effectiveness) and that efficiency gains would come with increased production speed. Among the authorities that have longer experience, the focus of implementation has moved towards increasing the complexity of the system, integrating data and procedures, and generally addressing strategic as well as operational needs.

Administrative operating departments increased their applications with either management functions (re-calculation) and planning functions, or with graphical presentation. Departments already working in a graphical way increased their applications with GIS analyses functions (like surface analysis). The evidence collected would indicate that relatively centralized top-down implementation leads to a wider and more co-ordinated diffusion within the organization than a bottom-up approach. The organizational consequences of GIS are very diverse and appear to be difficult to predict given the fact that half of the number of departments preparing GIS are not yet aware of the consequences.

Regarding the execution of tasks, automation appears to increase the quality of the final products and the number of applications (extension of analyses and new applications, including graphical presentation). All respondents reported a considerable increase in the use of the system and frequency of access. Regarding the quality of data and functions, almost all respondents agreed that both are improved by the implementation of GIS. GIS appear to improve the quality of the products (easier to design alternatives, make corrections, etc.) but do not reduce their costs. In fact the opportunities created by automation/GIS to design more alternatives, conduct more analyses, improve presentation and correct mistakes, all tend to increase the amount of work needed from users as well as the level of skills required. More work as well as greater skills needed are not always felt as an improvement by the respondents. The increasing demand for training may also result in conflicts with older employees.

Regarding the setting of the objectives for GIS in the future only the most ambitious larger municipalities will be able to execute operational processes as well as support decisions on a tactical and strategic level of planning by means of GIS. Most other authorities are likely therefore to focus their GIS on basic operational tasks.

FACTORS CONDITIONING THE DEVELOPMENT OF GRAPHICAL AUTOMATION

According to the findings of the research GIS are conditioned by three elements: the size of the organization, the experience with new technology and external circumstances.

Size of the municipality   The complexity of automation appears to be connected with the size of the municipality, which also influences the number of automated databases, networks in operation, and the way in which management information is being generated with automated applications. However, it is not true that automated information systems in the biggest municipalities are more integrated than in medium-sized municipalities. In fact, even in terms of costs of automation, developments and maintenance per inhabitant per year are more expensive in very large municipalities (>100 000 inhabitants) than they are in medium size municipalities (20 000–100 000 inhabitants) as shown in Figure 12.5.

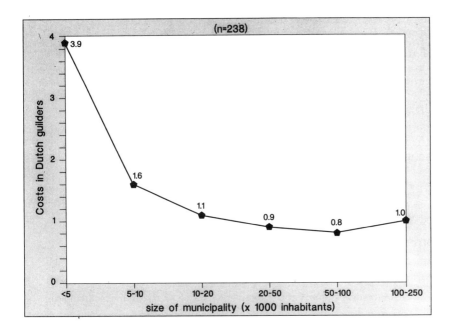

Figure 12.5 The average costs of one automated information system per inhabitant, per year, related to the size of the municipality.

Experience with new technology Larger Dutch municipalities started automating earlier than smaller ones. Experience with automation leads to more automated information systems, but not directly to integration of administrative/graphical data. The success of the introduction of municipal wide (integrated) GIS applications is independent of the experience or the size of the municipality. Even laggards and medium-sized municipalities can implement GIS successfully. On the other hand the experience with GIS leads to more complexity in another way, regarding the execution level (from supporting operational activities to strategic decisions) and the decision level (from registration to decision supporting).

External circumstances In general the development of GIS in the bigger municipalities in the Netherlands is not strongly affected by external circumstances. However, contextual factors play an important role among smaller authorities, and include the supply of the hardware and software, legislation and de facto standards set by very large organizations like the National Cadastre.

THE DEVELOPMENT OF GIS DESCRIBED IN A MODEL

The findings of the case-studies can be generalized in the model shown in Figure 12.6. According to this model the introduction of new technology (e.g. GIS technology) is conditioned by the size of the municipality and by external factors. The way municipalities experience the new technology depends on a number of

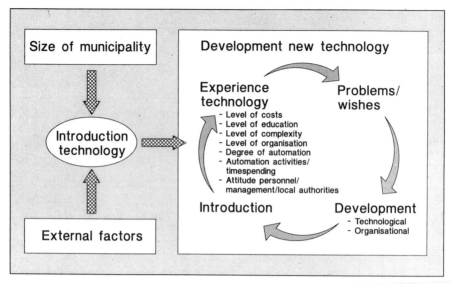

Figure 12.6   Model of the development of GIS automation, based on research in 428 municipalities in the Netherlands.

variables, such as:

- degree of automation;
- type of automation activities and time spent on automation;
- automation costs;
- level of skills available;
- degree of complexity of automation [extent for decision making, level of execution (strategic and operational), amount of integration, data communication and databases];
- hierarchical level of co-ordinating organization for graphical automation;
- attitude of personnel, management, and the organization as a whole towards automation.

This experience can be gathered in a relatively short time. Thus the pace of automation is not fixed, but bound by certain limits determined by the size of the municipality and external elements. It is not the starting date that settles the way in which automation develops, but the number of automated systems and their complexity. In the course of time, experience increases via a process shown in Figure 12.6 as a circle. The development process starts through the problems and perceived needs of the organization. These can be broken down into:

- technological problems and needs;
- organizational problems and needs;
- personnel problems and needs;
- budget problems and needs;
- problems and needs regarding product quality.

The choice of a mix of measures to solve these problems or to meet these needs depends on the degree of experience of the organization. During the start of automation, problems and needs in the field lead above all to technological development (more automation and more complexity) and in a later stage to organizational development (re-organization, change of hierarchical level of co-ordinating organization). In the beginning the development will be more technically driven, later on more determined by organizational issues. Through the introduction of these technological and/or organizational changes a new situation develops after which further changes can take place.

APPLICATION OF THE MODEL BY PHASING THE DEVELOPMENT
OF INFORMATION SYSTEMS

This section describes a normative phase model for the development of (graphical) automation in municipalities. The model adapts those proposed by Nolan (1982) and King and Kraemer (1984) on the basis of the Dutch research findings. It may be assumed that the phasing proposed will also apply to municipalities in other countries and to other similar data processing parts of government services, like provinces or regional co-operative bodies and ministries of the national government.

The model is based on two essential elements:

1 The interaction between the supply of the hardware and software and the organization's demands (problems and needs). This interaction can be explained by theories about technological and economic determinism or social interactionism (see Chapter 3 in this volume);

2 The organizational level, which determines the development of automation.

The change in attitude of top management from minor commitment into real commitment is the turning point for the development of automation. Before the commitment, the development can be characterized in particular as being bottom-up. Thereafter the development is controlled more and more from the top-down, directed towards strategic information systems. The top-down controls change the character of the automation process, including the development of applications, the organization of automation and orientation of the automation department and the involvement of users.

With respect to applications, bottom-up developments tend to support objectives at a low organizational level even though in some instances links between applications are introduced. Top-down development instead directs towards an infrastructure, including information architecture, technical infrastructure and physical data structure which best supports the objectives of the organization at macro-level.

With respect to the organization of automation, bottom-up developments tend to focus on technological support for local-oriented data administration. Top-down developments involve infrastructure development with particular attention for standards, multi-used databases (so called basic registrations), communication networks and meta data.

With respect to user involvement, during the bottom-up development users are especially directed towards the realization of their own objectives at a low organiz-ational level. This attitude continues to exist during the top-down development; they

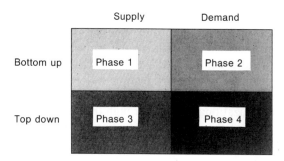

Figure 12.7 Phasing development of automation in local authorities based on technical and organizational determinism.

accept the imposed restrictions of their freedom on behalf of the infrastructure development of which they recognize the advantages more and more providing that the realization of their own objectives is not endangered. Although the attitude of the top-management is essential, the development of this attitude is not an autonomous process, but rather the result of the developments of users. Due to the fact that information need and involvement of users change, its service and the need for control develop as well. Automation and infrastructure development are not objectives in themselves, but means to achieve a better operational management. There will always be data processes which will function better without automation.

Figure 12.7 shows the relation between supply/demand and bottom-up/top-down directed automation. The matrix shows four phases. On the basis of this matrix and the above described developments the following classification of phases can be composed (see Table 12.1). Phases 1 and 3 focus more on the supply side, while, in phases 2 and 4 the emphasis shifts to the demand aspect. The introduction of new technology takes place in phase 1. If the technology finds a positive environment, it can spread quickly to many departments in the organization. Phase 2 is a phase which shows at first sight two sorts of conflicting developments. On one hand the individual automation continues, on the other hand attempts are made to introduce standards in order to prevent the creation of more information islands. During these attempts those departments which can be considered as pioneers of the automation play an especially important role because their support for standardization is critical. In phase 3, the emphasis is on infrastructure development. Decisions with organizational consequences are made and the co-ordination of the information systems is raized to a higher hierarchic level. Integration of the infrastructure takes place in phase 4.

During the first two phases bottom-up growth (from the departments) is encouraged. Phases 3 and 4 need control from the management (top-down). According to the evolutionary theory (Greiner, 1972; Nolan, 1979) every phase would end in a crisis. Subsequently the solution to end this crisis would be the characteristic of the next phase. In the above mentioned phase model these crises are formulated as follows:

end phase 1 pioneers realize that further development is slowed down by the remaining departments;

Table 12.1   Phasing of the development of municipal (graphical) LIS automation

Phases	General characteristics
1 Initiation	Development of automated applications, enthusiasm, information islands and big differences between departments
2 Local control	Development of standards, more complex applications and first (internal) data communication, first supporters of infrastructure (basic registrations) and first initiative to that, attempts are made to involve remaining departments in a certain way, a few stay entirely behind, management gets more and more involved and aims at formalization of decisions about information systems
3 Infrastructure development	First basic registrations, infrastructure development gains support, control becomes more top-down, substantial change in organization with co-ordination at a higher level
4 Integration	Adjustment applications to infrastructure (basic registration), integrated (graphical) information systems, organization wide GIS

end phase 2 top-management realizes that the present automation development becomes uncontrollable and recognizes the strategic value of information systems;

end phase 3 excessive formalization which limits the freedom for innovator ideas.

Differentiated Phase Models for Automated Data Processing

To apply this general municipal phasing mode, it is useful to differentiate phasing by type of automated system being introduced (GIS, administrative data processing and technical information system management) and by size of municipality.

The division between GIS, administrative data processing and technical support functions respectively is connected with the usual division in Dutch municipalities between a surveying department, which manages and distributes topographical data, the administrative operating departments (in which the tax department is of great importance) and the IT department providing technical support.

Phase models for automated administrative and graphical LIS

Table 12.2 shows the phasing of the automated administrative data processing. A distinction can be made between a part which will incorporate graphical presentation and analysis by means of GIS, as for instance environmental management, and a part which will continue to be only administrative, like tax registration. Table 12.3 shows the phasing of GIS.

The history of development is of great importance during the phasing of technical system management. A municipality which has built up a computing centre in the past and has purchased a mainframe, will show quite another development of system

Table 12.2 Phasing development of municipal automated administrative data processing

Phases automated administrative data processing	Characteristics of the part that will function only in an administrative way	Characteristics of the part that will develop into GIS
1 Initiation	Registrational applications	
2 Local control	Automation of more complex functionalities (analyses, planning, management)	First graphical presentation
3 Infrastructure development	Applications using basic registrations	Integration with topographical data
4 Integration	Analyses functions and decision support	Use of GIS, analyses functions and decision support

Table 12.3 Phasing development of GIS in municipalities

Phases automated graphical geo-data processing	Characteristics
1 Initiation	Automated map production
2 Local control	Object-oriented data structure, first and local link between administrative and graphical data
3 Infrastructure development	Organization wide integration administrative/graphical data
4 Integration	Organization wide integrated GIS applications used for decision support

management as, for instance, a municipality which used to contract out its automation in the past and which started automation in-house only recently. In the former a centralized system management is more likely to retain control, while in the latter the distribution of PCs may lead to a decentralized system management. Bearing this in mind, a phasing starting with the introduction of personal and mini-computers and developing by means of linking PC-networks and mini-computers towards the introduction of organization wide communication networks with matching standards is quite conceivable.

Phase models differentiated on the basis of municipal size

For the purpose of this phasing, four municipal sizes can distinguished in the Netherlands, namely: small < 40 000 inhabitants, medium-sized 40 000–100 000, large 100 000–225 000 and very large > 225 000 which includes the four largest municipalities. It may be assumed that the end phase of automation for municipalities within each of these categories may vary (i.e. some will go all the way to phase 4, integration, while others will stop earlier). Realization of an organization wide integrated GIS depends on the following circumstances:

1 The 'importance' of an integrated system for internal working processes: this increases with the size of the organization. Data flows between departments and coordination tasks increase exponentially with the size of the organization.

2 The 'technical feasibility' of an integrated system, depending on the existing equipment and software and the available know-how in the organization concerned: this increases with the size of the organization and/or the size of the autonomous units. Bigger organizations have more know-how concerning automation at their disposal.

3 The necessary effort for the 'organizational feasibility' of an integrated system. Organization wide re-organizations, including the creation of support for changes and the possibilities for mutual tuning, will require more effort in bigger organizations than in small organizations.

Subsequently it may be assumed that in those organizations where the importance and/or the feasibility are minor, integration is not likely to be implemented. Therefore it is uncertain whether a fully integrated GIS will be realized in small municipalities. For small municipalities this implies that (without special intervention) phase 2 will probably be the end phase, in which automated applications will be partly linked, having the possibility to support more complex functions. For a part of these municipalities it remains to be seen whether they will start GIS. This will depend on the cost development, the user friendliness of the equipment and costs of digital topographic data, which will be available for a large part of these municipalities in the Netherlands within a few years.

Medium-sized and bigger municipalities will (or want to) grow towards phase 4. However, it remains to be seen whether the bigger municipalities will introduce phase 4 completely and across the municipality. Representatives of the biggest municipalities in the Netherlands state that the introduction of organization wide integrated systems in municipalities of more than 150 000 inhabitants is not very likely. For sure, these municipalities will develop complex automated applications with the possibility of supporting management decisions. It is expected that the medium-sized municipalities are most likely to carry through organization wide integrated GIS. It will take a long time before this phase will be realized. Starting from 1995 a period from 10 to 20 years has to be taken into account. Further development of technology will be of great importance.

CONCLUSIONS

This chapter has traced some of the background to the diffusion of GIS in Dutch local government and presented a model for the development of graphical information systems. Although some indication of the possible phasing based on size of population has been provided, it must be recognized that external factors may influence phasing considerably. For example, statutory obligations for the introduction of integrated information systems can alter the phasing substantially. In this case small municipalities will introduce infrastructure development themselves or contract it out. The most important differences between various countries and organizations may lie in these external factors. The development of the supply and the prices of computer hardware and software in combination with the economic situation will influence the phasing as well. If information systems get cheaper and more simple to operate and digital data become more easily available and cheaper, barriers to introduce integrated systems will lower as well. Nevertheless organizational issues will continue to play a key role for the diffusion and effective utilization of automated systems like GIS as confirmed in the other chapters of this book.

REFERENCES

BENBASAT, I. *et al.*,, 1984. A critique of the stage hypothesis: theory and empirical evidence, *Communications of the Association for Computing Machinery*, **25**(5), 476–85.

GRAAFLAND, A., 1989. Analysis of municipal information supply, *Proceedings of the 13th Urban Data Management Symposium*, Lisbon, May 29–June 2, pp. 293–305, Delft: Urban Data Management Society.

GRAAFLAND, A., 1991. Implementation of information systems in municipalities in the Netherlands *Proceedings of the 1st International Conference on Municipal Information Systems MIS'91*, Prague, November 11–15, pp. 31–8, Prague: Institute for Municipal Informatics Zdiby: Research Institute of Geodesy, Topography and Cartography.

GRAAFLAND, A., 1992. Implementation of graphical information systems in municipalities in the Netherlands, *Proceedings of the 15th Urban Data Management Symposium*, Lyon, November 16–20, pp. 499–509. Delft: Urban Data Management Society.

GREINER E.L., 1972. Evolution and revolution as organizations grow, *Harvard Business Review*, July–August, 37–41.

KING, J.L. and KRAEMER K.L., 1984. Evolution of organizational information systems; an assessment of Nolan's stage model, in *Communications of the Association for Computing Machinery*, **25**(5), 466–75.

KING, J.L. and KRAEMER K.L., 1991. Patterns of success in municipal information systems: lessons from US experience, in *Informatization and the Public Sector*, *Vol. 1*, (1), pp. 21–40. Amsterdam: Elsevier Science Publishers BV.

KRAEMER, K.L. and KING, J.L., 1981. Computing policies and problems, a stage theory approach, in *Telecommunications Policy*, September, pp., 198–215, Guildford: IPC Science and Technology Press.

NOLAN, R.L., 1979. Managing the crisis in data processing, *Harvard Business Review*, March/April, 115–26.

NOLAN, R.L., 1982. *Managing the Data Resource Function* St Paul, Minnesota: West.

PART IV

Evaluation

A comparative evaluation of GIS diffusion in local government in nine European countries

IAN MASSER and MASSIMO CRAGLIA

INTRODUCTION

The nine country case studies contained in the two preceding sections of this book present a variety of perspectives on the diffusion of GIS in local government in Europe. They also highlight the diversity of arrangements that have come into being to carry out the responsibilities of local government in these countries and the ways in which specific national circumstances influence the diffusion of GIS in particular cases. Despite these differences, however, the findings of the case studies have many common features and there are some striking similarities in the experiences of countries which otherwise appear to be very different.

With these considerations in mind this chapter evaluates the findings of the nine country case studies from a cross national comparative perspective. The process of comparative evaluation is divided into three separate stages. The first of these involves the construction of a profile for each country which summarizes the distinctive features of national experience. These features are then compared with those of the other eight countries within a common analytical framework at the second stage, while the third and final stage explores the extent to which a typology can be developed on the basis of the differences observed between countries.

THE NINE NATIONAL PROFILES OF GIS IN LOCAL GOVERNMENT

Great Britain

In many respects the two surveys analysed by Ian Masser and Heather Campbell in their contribution represent the benchmark for the other four national case studies in Part II of the book. Their complete coverage of all local authorities, and their sequence in time also give some useful insights into the dynamics of the GIS diffusion process.

Great Britain stands apart from other European countries for the large size of its lower tier local authorities, the smallest district having some 24 000 people, and the availability of large scale digital base maps for all urban areas from Ordnance Survey (OS). The human and financial resources at the disposal of these relatively large authorities and the increasing availability of up-to-date digital data provide the background to the recent diffusion of GIS in British local government. The findings of the 1993 survey indicate that 30% of all 514 local authorities had GIS. Although uncertainty about the reorganization of local government and financial constraints are holding some authorities back from investing in GIS, Masser and Campbell argue that the Service Level Agreement reached between the OS and local authorities in 1993 is likely to provide a significant boost to further GIS diffusion. The authors also highlight the different patterns of GIS diffusion in Great Britain which follow closely the hierarchy of local government (88% diffusion among counties and regions, 49% among metropolitan districts, 17% in shire and Scottish districts) as well as the core-periphery model of diffusion given that the probability of a local authority having GIS is higher in the wealthier South of the country than it is in the North and Scotland, particularly among the smaller authorities.

The extent to which GIS is a relatively new phenomenon even in British local government is underlined by the finding that three quarters of all systems have been purchased since 1990 and that the average length of experience with GIS is slightly less than 3 years. Small authorities tend to have only one GIS system and to use it in a more collaborative inter-departmental fashion than larger authorities where it is common to find more than one independent departmental system.

Planning and development departments are the leaders in the adoption of GIS (one quarter of the total) followed by the highways and estates departments which together account for another quarter of all departments involved in GIS. IT and technical services are also well represented as are legal services, parks and recreation and the Chief Executive's department.

The overall finding is therefore that GIS in British local government is largely decentralized and bottom-up in nature. The predominance of planning as the key department in the diffusion of GIS reflects the familiarity of planners with geographic information as well as the traditional responsibility of that department for meeting the mapping needs of the authority.

In terms of software, the key finding is the wide range of GIS software in use, almost 40 different packages. Arc/Info is the market leader (22%) particularly in larger authorities (1 in 3). By contrast, in small authorities Arc/Info accounts for only a 10% share, the emphasis here being on British products [e.g. Axis, Alper Records (Sysdeco), G-GP] which account for almost one third of the systems in use.

The main perceived benefits of GIS cut across authority type and are largely confined to improving information processing capabilities (60%) followed by better decisions (21%). Only some 11% of respondents identified savings of any kind as the main benefit. The type of problems experienced are almost evenly spread among the three main categories, technical, organizational and data related, and include problems of hardware reliability, lack of skills and appropriate organizational arrangements, as well as concerns over the costs of data capture.

Comparing these 1993 results with those of the 1991 survey underlines the rapidity of GIS diffusion in British local government. During the $2\frac{1}{2}$ years between the surveys the number of systems in local government has effectively doubled and the number of authorities with GIS has increased by 75%. During this period there have also been rapid changes in technology with the almost complete demise of main-frame installations. Interestingly, these rapid changes have not been translated into significant differences in the problems and benefits of GIS perceived by the users.

Germany

The contribution from Hartwig Junius and his colleagues at Dortmund draws attention to the slow progress that is being made in the computerization of the Cadastre in Germany and the need for greater central co-ordination of geographic information handling activities to establish common principles for spatial referencing. The proposals put forward by the Association for German Cities for the development of the spatial reference system MERKIS are particularly interesting as an attempt to integrate local government databases and avoid the duplication of data capture and maintenance. However, the problems of implementing these proposals within existing organizational structures are considerable as the case study of Wuppertal indicates. This is confirmed by the findings of the survey conducted by Junius et al. which shows that only one third of all potential applications follow the MERKIS recommendation.

For practical reasons the survey was limited to the 86 German cities with populations of over 100 000. There is no information therefore available for the 14 273 smaller municipalities on the bottom tier of local government or the 457 non-metropolitan counties that form the second tier within the federal structure. However, it is safe to assume that to date only a few of the smaller municipalities have GIS experience.

The findings for the large cities indicate the widespread nature of GIS diffusion at this level of local government in Germany. Of the 80 cities who responded to the survey, 70 had already adopted GIS, and the remaining 10 were in the process of acquiring facilities. However, only 45 of these authorities had full-scale integrated GIS facilities, indicating that the diffusion curve is still in mid-phase.

Given the emphasis placed on land information systems in Germany it is not surprising that GIS diffusion in local government has been fastest in surveying departments where GIS is routinely applied to daily work in 70% of the cities with GIS facilities. Other departments lag some way behind the surveyors in this respect with only one in eight planning departments routinely using GIS in their daily work.

A similar picture emerges in terms of spatial database development and the choice of software. Spatial databases are under development in all cities for surveying applications and database development for planning and utility applications is taking place in only a limited number of cities. Both the two clear software leaders in terms of market share, SICAD by Siemens-Nixdorf and the ALK-GIAP software developed by AED in Bonn are primarily tailored to surveying applications. In contrast, only four of the cities surveyed used Arc/Info.

The authors conclude that GIS diffusion in Germany has been handicapped by the federal organization which gives a relatively large degree of autonomy to local government and provides little incentive for collaboration. Nevertheless, the findings of the survey indicate that there has been a high level of GIS adoption in the main cities although GIS is largely used for surveying applications.

Italy

Luisa Ciancarella and her colleagues explore the diffusion of GIS in Italian local government at a time of political upheaval when considerable changes are taking place to the traditional political cultures of both central and local government. These changes coincide with a general tendency towards greater decentralization following the 1990 reform of local government which considerably extended the powers and autonomy of the provincial and communal tiers of local government. Under these circumstances it is not surprising to find that the patterns of GIS diffusion described by the authors are quite complex.

In all, two thirds of the regions already have GIS. There are close links in most cases between the adoption of GIS and the development of digital map bases for these regions given the slow progress that has been made in this respect by the national mapping agency and the Cadastre.

The regions have pursued a variety of strategies with respect to the lower tier authorities. Some have paid for the full cost of provincial GIS development provided that these authorities use the same software and data standards. Others have developed pilot projects at the municipal level. The impact of these strategies varies widely from region to region. In Emilia Romagna, for example, all nine provinces have GIS which reflects the strong planning traditions of this region. Elsewhere, the impact of these strategies and similar central government initiatives has been very limited.

Nevertheless, over a third of the provinces already have GIS and this proportion is increasing rapidly (one in three have firm plans) as a result of the additional responsibilities given to them by the 1990 reforms. This is particularly important with respect to the requirement to prepare provincial land use plans. Within these provinces, there is also a very strong bottom up dimension to GIS adoption which appears to be particularly effective in promoting implementation.

The position of GIS at the municipality level generally reflects the low level of computing capabilities in the 8100 communes in Italy. This is largely a function of population size as 87% of the communes have populations of less than 10 000. However, by comparison with the regions and provinces, the diffusion of GIS is relatively limited even in the medium and larger cities which are in a position to command IT resources. The findings suggest that less than 10% of all municipalities in the 50 000–250 000 range had GIS and the proportion of GIS in the larger cities was even lower. A key factor with respect to the latter may be the continuing uncertainty surrounding the establishment of 12 new metropolitan authorities which were introduced in the 1990 reforms.

The dominant motivating factors for the acquisition of GIS in Italian local government are digital map production and the implementation of land use planning tasks. The software market leader in the regions and provinces is Arc/Info with over half the total applications. The impact of this package has been much less marked in municipal applications where there is greater diversity.

There is a very clear North South divide in Italy with respect to GIS diffusion among regions and municipalities as most GIS applications are located in the North of Italy. This divide is less dominant among provinces where 50% have GIS in the Centre-North and 20% have GIS in the Centre-South.

The findings of the surveys of the regions and provinces show that the most important perceived benefits of GIS were improved information processing particularly in the fields of automated mapping production and thematic mapping. The predominant problems cited were associated with organizational issues. These include lack of coordination, lack of skilled staff and a general lack of awareness.

In overall terms, recent political upheavals in Italy have both constrained the diffusion of GIS in the large cities whilst at the same time creating new opportunities for the provinces. The slow progress that has been made by the national mapping agencies and the Cadastre in developing digital topographic data bases has meant that local authorities together with other key users of geographic information in the utilities fields have had to spend a considerable amount of their time and resources in developing their own data bases for planning purposes. The overall effect of this on GIS implementation at the municipal level has been considerable. However, it should be noted, even at this level, that there are a number of examples of municipal authorities which appear to have innovative approaches to the development of GIS applications.

Portugal

Antonio Arnaud and his colleagues present a fascinating account of GIS on the point of very rapid take off due to a series of proactive government initiatives to modernize Portuguese local government. They include the creation of a National Centre for Geographic Information (CNIG) under the auspices of the Ministry of Planning and Territorial Administration to oversee the development of a national geographic information system and recent legislation requiring all 305 municipalities in the country to prepare municipal masterplans. In February 1994 a major boost to the use of GIS in local government was provided by two government measures providing financial support for the computerized management of municipal masterplans and the development of local nodes for the national geographic information system respectively.

The findings of the postal survey of all municipalities carried out by the authors in late 1993 and early 1994 therefore provide a useful account of GIS in Portuguese local government prior to the implementation of these financial incentives. They show that adoption levels in general were low, reflecting the limited experience of IT in most Portuguese municipalities. At the time of the survey only 12 GIS facilities were in operation or still under development and a further 24 municipalities had some form of AM/FM facilities. Half the GIS and a third of the AM/FM facilities were in urban authorities. A good example of these pioneer GIS authorities is the city of Oeiras where strong local leadership coupled with clearly defined planning objectives and the restructuring of its administrative organization enabled the municipality to acquire substantial financial support for the development of its GIS facilities, being the first to apply for EU funds managed by the Regional Commission of Lisbon and the Tagus Valley.

The findings of the survey also indicate a high overall level of awareness of the potential of GIS among municipalities which reflects the impact of recent government initiatives. Of those municipalities which are currently without GIS or AM facilities and which responded to the questionnaire, three quarters planned the introduction of GIS or AM facilities and only a relatively small proportion of mainly rural municipalities stated that they had no plans for GIS. If these expectations are fulfilled throughout the country the overall take up of GIS and AM/FM could rise from around 12% to nearly 90% of municipalities over the next few years.

As might be expected under these circumstances the range of applications cited by the respondents reflects intentions rather than experience. Nevertheless they show a strong planning bias reflecting the impacts of recent legislation requiring all municipalities to prepare municipal masterplans and regulating the role of the municipality in managing its territory.

Parallel to the questionnaire survey the authors carried out a number of interviews with municipalities to explore some of the problems associated with the introduction of geographic information technology in local government. Human resource issues identified by these means include the lack of local champions and the need to foster greater computer awareness amongst staff. Organizational barriers to GIS implementation include internal bureaucratic structures, lack of skilled staff and continuing technical support from vendors.

Denmark

According to Hans Kiib, Denmark is one of the most decentralized countries in Europe with respect to the powers that have been devolved to its municipalities and counties. This is particularly evident in the economic base as between 70 and 80% of local government income is derived from municipal taxes and the level of local government employment as a percentage of the total national labour force is 16.8%. The municipalities are also the main providers of large scale maps. Together with the Danish Survey and Cadastre they are co-managers of the Cadastre and they also maintain comprehensive register information relating the land parcels, property, zoning and taxation.

Since the early 1970s there has been extensive use of IT in Danish local government. In 1972 Kommune Data was set up as an intermunicipal enterprise to provide technical support for the development of a wide range of register based information systems associated with local authority management. In the early 1980s there was a shift from manual cartography to digital mapping within local government and the need for a coherent national land information strategy is now widely recognized.

Given these circumstances, it is not surprising to find that more than 80% of all municipalities maintain register based information systems where georeferencing is present. The findings of a comprehensive telephone survey conducted at the beginning of 1993 states that either GIS or AM/FM technology is almost universal in municipalities with populations of over 50 000. Levels of adoption are generally lower in the smaller municipalities although many of these make extensive use of CAD systems. The vast majority of GIS facilities have been acquired since 1990 and nearly half the municipalities with less then 25 000 population are planning to purchase GIS or AM/FM facilities by the end of 1994.

GIS and AM/FM facilities are used primarily for digital mapping in most authorities but they are also widely used for utility management. Up to 1993 there had been no technical and organizational integration between the register information and the digital large scale maps. For that reason GIS or AM/FM has not been used so far for planning and land management in local government.

At the time of the survey 80% of applications relied on two Danish and two north American software packages: Dangraf developed by Kommune Data and Jydsk Telephone, GeoCad, Intergraph MGE and AutoCad. At the time of the survey the market share of Arc/Info was restricted due largely to the lack of a Danish vendor.

The findings of the survey generally show that the level of GIS and AM/FM adoption in the counties is lower than that for the larger cities. This essentially reflects the distribution of responsibilities between the two levels of local government in Denmark which allocates land information functions including zoning and taxation to the municipalities and hospitals, health care, public transport and rural planning to the counties.

Greece

According to Dimitris Assimakopoulos the current state of geographic information systems technology in Greece is less well developed than in countries such as Great Britain and the Netherlands because of an overall lack of resources. Nevertheless the Greek GIS community is growing rapidly around a core of 60 key members working in a variety of administrative settings. A major factor in promoting the diffusion of GIS has been the availability of funding from the European Union. It has been estimated, for example, that the European Community Social Fund (ECSF) has played a major role in the development of GIS in between 30 and 40 of the 361 cities in Greece. Despite the impetus provided by such EU initiatives, the diffusion of GIS in Greece has been hampered by the lack of basic resources. This is particularly evident with respect to digital data. The Hellenic Military Geographic Service has released 5000 scale contour data for some city councils in the Athens and Thessaloniki areas and roads and administrative boundaries data are available only at the 1 million scale for the country as a whole. In addition to this lack of data there is also a lack of skilled manpower resources particularly in local government. To deal with these circumstances Greek municipalities have contracted out part or all of their IT applications to private sector organizations and university laboratories. These organizations have also sought to fill some of the gaps in data by developing their own products.

Municipal applications are confined without exception to the larger urban authorities as the average population of the Greek parish is around 1800 people and the scope for GIS applications at the prefectural level so far has been limited. Nevertheless, as the case studies of Volos and Kos show, Greek local government contains some interesting examples of innovative cultures that have sought to exploit the opportunities opened up by GIS. A number of interpersonal communication networks have also come into being to facilitate the exchange of ideas and experience over the last few years. A key feature of these networks is the extent to which they link like minded people with similar professional backgrounds in surveying and engineering.

France

Philippe Miellet's contribution highlights the extent to which recent developments in GIS build upon a long tradition in geographic information handling in many of France's larger cities. This goes back to the 1970s when the first urban information systems (BDUs) were set up in Paris, Lyon and Marseille. Consequently recent efforts to develop GIS have been able to draw upon a considerable body of operational and organizational experience in handling geographic information.

Miellet's analysis also demonstrates the importance of digital data provision in the diffusion of GIS. Small scale data provision in France is in the hands of the national mapping agency the Institut Geographique National. Its main products include basic cartographic and topographic data at the 25 000 scale and above. Its handling of topographic data have been particularly innovative. Each object is captured in three dimensions. Therefore complete spot height information is available for the whole country. Large scale data provision in France is closely connected to the maintenance of the Cadastre. This consists of 560 000 map sheets containing information about 100 million parcels of land. Responsibility for the Cadastre lies in the hands of the National Tax Office. Although attribute data has been available in digital form since 1990 it is estimated that it will take 30 years to complete a fully digital Cadastre even with the help of the local authorities.

There are about 36 000 communes representing the lowest level of government in France with an average population of about 1500. As might be expected the diffusion of GIS is closely linked to population size. Most of the urban authorities with populations of over 100 000 had GIS by 1993 whereas very few authorities with less than 20 000 had GIS. A distinctive feature of most urban GIS is their interdepartmental and intercommunal nature: the former a consequence of the central role played by the digital Cadastre in geographic information handling, whereas the latter reflects the large number of intercommunal organizational structures that have come into being to ensure the delivery of basic urban services. Of particular importance in this respect are the Urban Communities that were set up in the early 1980s for the management of most of the larger cities.

In contrast, GIS diffusion at both the department and regional levels is much less evident than in the larger cities. There is also only a very limited history of geographic information handling in these agencies and most of the applications that have been developed are limited rather than comprehensive in scope.

Poland

In their contribution, Malgorzata Bartnicka, Slawomir Bartnicki and Marek Kupiszewski present some fascinating insights into the diffusion strategies for GIS in what until recently was a centrally planned economy and which has undertaken the conversion to market economy faster than other Eastern European countries. The context of this chapter is therefore one of rapid change in which old and new coexist and fight for supremacy.

Local administration is divided into regional, sub-regional, and local levels with central administration retaining a large degree of control. The authors underline that a clear division of responsibilities between these levels has yet to emerge as previous governments attempted to decentralize the system, while more recently the process has been reversed. The dual system of centrally controlled and locally determined administration is reflected in the existing processes of GIS diffusion in local government analysed by the authors: the central model and the local model. The central, top-down model of GIS diffusion is based on a hierarchical structure of state agencies acquiring, and retaining the control and ownership of all spatial data. Whilst the apparent high degree of awareness of GIS by the Polish central government emerges from the discussion, the extent to which this technology is perceived as a key 'element of the organizational infrastructure of the State' is a healthy reminder of the potential implications of highly centralized GIS so clearly illustrated in the chapter by Wegener and Masser.

In operational terms, the enormous financial, technical and organizational diffi-culties faced in implementing the central model are clearly shown by the authors through the case study of the pilot project established for the region of Lodz. The authors argue that many of these difficulties could be overcome by greater dialogue and partnership with local users, and support this contention with three case studies of locally based bottom-up approaches in the region of Plock, and in the towns of Sierpc and Szczecin. In contrast to the highly bureaucratic central model, these local case studies are striking for the pragmatism of users, who are directly responsible for the choices made in implementation, and for the partnerships established with local utility companies and/or the private sector to share costs and expertise. In discussing these case studies, the authors argue that the availability of technology is no longer a key problem in Poland, although in a broader perspective there are still numerous limitations with respect to the development of supporting infrastructure and technical services. In their view, data are the critical issue not because of lack of data (although this may also be the case in some instances) but because data often need to be updated, structured, digitized and integrated. In this respect, the authors argue that the local model of GIS diffusion appears to be much more effective because only the data that are really needed are captured and maintained, while the central model is too ambitious and inflexible.

In summing up, the authors identify advantages and disadvantages of both approaches and argue that the local model of implementation is more likely to satisfy the increasing demand for GIS in Poland in the coming years. The extent to which this latent demand at the local level is supported by available skills in handling geographic data remains open to question.

Netherlands

In his chapter, Ad Graafland provides a broad overview of the diffusion of GIS and graphical automated systems among Dutch municipalities. The system of local government in the Netherlands is partly centralized and partly decentralized with both central government and local authorities (650 municipalities and approximately 100 'polder-boards') having strong powers while the middle layer of 12 provinces is somewhat weaker with less clearly-defined responsibilities. Graafland indicates that Dutch local government as a whole has a well established IT base. All municipalities however small appear to have access to IT, and although there are obvious variations in skills and resources available according to population size, most municipalities over 50 000 people and about half of those over 20 000 have GIS. Availability of large scale digital maps appear to be a significant barrier to the diffusion of GIS as the National Digital Cadastral programme has made only slow progress. Nevertheless, the bottom-up approaches of individual municipalities appears to be complementing national scale activities even if at the expense of coordination and standardization.

Graafland focused his research on the diffusion of GIS within organizations of medium/large municipalities amongst which he conducted a number of case studies. Based on the findings of this research he argues that the diffusion of GIS is conditioned by the size of the organization, external circumstances, and experience with technology. Whilst there is evidence that larger authorities tend to have longer experiences with IT and more complex systems in place, the research confirms that this is no guarantee of greater integration particularly between alphanumeric data for management and administration and graphical data for survey and planning. Therefore smaller organizations starting automation later are just as likely, if not more so, to be able to implement more integrated IT and GIS environments.

COMPARATIVE EVALUATION

The main features of GIS diffusion in local government in the nine countries that are described above are summarized in Tables 13.1, 13.2 and 13.3. Table 13.1 compares the characteristics of these countries with respect to the institutional context within which GIS diffusion takes place in local government, the structure of local government and extent of digital data availability.

Institutional Context

From Table 13.1 it can be seen that there are important differences between the nine countries with respect to their political and economic stability. Recent political upheavals in Italy, for example, have profoundly affected the political culture that underlies both central and local government in that country. Poland is in the middle of a major structural transformation from a command to a market economy. Germany is also undergoing important structural changes as a result of the incorporation of the East German *Länder* into the former West German administrative system. In contrast, Denmark, France and the Netherlands have relatively high degrees of political and social stability which is also reflected in local government.

In several of these relatively stable countries, however, there have been moves to decentralize powers from central to local government over the last two decades. The impacts of decentralization are most pronounced in Denmark following the reforms of the 1970s, but similar trends can be found in France and Italy during the 1980s and more recently in Greece and Portugal. In Great Britain developments during the last decade and a half run counter to this general trend with more central government control over local government. Poland is also a country where central government control over local government is increasing once again after the decentralization of power following the collapse of the former communist regime.

These changes have given rise to uncertainties about the responsibilities allocated to various layers of local government in several countries. In Great Britain, for example, this culminated in the abolition of the metropolitan counties by the government in 1986. Similarly, the provinces in the Netherlands are struggling to find a role after changes in local government structure. In France, on the other hand, decentralization has given new powers to the departments as it has to the provinces in Italy.

There are also important differences between countries with respect to the relationship between local government and the private sector. In Greece, Italy and Poland, for example, contracting out of data collection and IT tasks to the private sector is commonplace because of the limited resources at the disposal of local government. In Great Britain, local authorities are also being increasingly obliged by central government to put key tasks out to competitive tender to reduce costs.

In Portugal, on the other hand, central government is taking an increasingly proactive role to modernize local government. Like Greece, the diffusion of GIS in this country has been substantially accelerated by the availability of funds from the European Union.

Structure of Local Government

Table 13.1 shows only the tip of the iceberg in that it refers essentially to the number of authorities at various levels of local government in the nine countries and gives some indication of the size of these units in terms of population. As such it does not deal explicitly with the great diversity that also exists with respect to the ways in which the various functions are discharged by the different tiers of local government in the nine countries.

Nevertheless, Table 13.1 highlights some important differences between the countries. Germany is the only country surveyed which has a federal system of government. France, Italy and Poland have developed three tier local government structures whereas all the other countries have two tier structures with the partial exception of Great Britain which has single tier authorities in the main centres of population since the abolition of the metropolitan counties. Great Britain also stands out from all the other countries with respect to the size of its lower tier authorities which, with relatively few exceptions, contain populations of over 100 000. Elsewhere in Europe, small is generally beautiful. Over 80% of Italian communes have less than 10 000 population. In France the average population per commune is 1500 and the comparable figure for Greece is 1800. Most German authorities have populations of less than 10 000 and even in Denmark, where there was a major reorganization of local government during the 1970s which resulted in a substantial reduction in the number of authorities, half the municipalities have populations of less than 10 000.

Digital Data Availability

Once again Great Britain stands out from all the other countries with respect to digital data availability given that its national mapping agency, the Ordnance Survey, provides a comprehensive mapping service for both large and small scale maps. Digitization of the former is complete and local government access to this information has been substantially facilitated by the service level agreement for the purchase of digital data reached between the local authority associations and the Ordnance Survey in March 1993.

At the other end of the spectrum comes Denmark which also has a very high level of digital data availability but makes the municipalities themselves the providers of large scale maps. These authorities also maintain the Cadastre together with the Land Registry as well as a large number of register based systems.

In terms of digital data availability Greece, Italy and Portugal in particular currently suffer from a proliferation of digital data sources as a result of the limited progress made by their central government agencies with respect to the provision of digital data.

Table 13.1 Key contextual factors in nine European countries

	Great Britain	Germany	Italy	Portugal	Denmark
Institutional context	Increasing pressures towards privatization of local government tasks has effected GIS diffusion. Uncertain impacts of ongoing review of local government in terms of the number and functions of authorities	Stable and well established local government culture. Incorporation of former East German *Länder* presents new opportunities	The 1990 reform of local government which extends powers and degree of autonomy of lower tiers has both constrained and created opportunities for GIS diffusion. Uncertain impacts of recent political upheavals on local government. Limited use of IT in local government	Proactive Government measures to modernize local government. Establishment of a National Centre for Geographic Information to coordinate the implementation of a national geographic information system. Availability of EC funding to support modernization	Widespread decentralization of powers to local government since 1970. Extensive use of IT in local government since the 1970s
Structure of local government	Two tiers: 47 shire counties and 9 Scottish regions with 333 shire districts and 53 Scottish districts. Single tier of 69 metropolitan districts. Few districts have populations less than 100 000	Three tiers: 16 *Länder*, 543 counties and large cities and 14 809 municipalities. Most of municipalities have less than 10 000 population	Three tiers: 20 regions, 103 provinces and 8100 communes; 87% of communes have less than 10 000 population; 12 new Metropolitan Authorities	Three tiers: 7 regions (including the Azores and Madeira), 29 districts and 305 municipalities. Average size of municipality 34 000	Two tiers: 14 counties and 275 municipalities. Half the municipalities have less than 10 000 population
Digital data availability	Comprehensive digital topographic data service provided by Ordnance Survey. Service Level Agreement reached with local authorities in March 1993 boosted GIS diffusion	Strong surveying traditions and collaboration between local authorities with respect to the development and maintenance of digital topographic databases (e.g. ALK and MERKIS)	Slow progress made by National Mapping Agency and Cadastre. Nearly half the regions developing their own digital topographic data bases. Profusion of data sources	Diversity of sources; 65% of country covered by 10 000 scale orthophoto maps from the National Mapping Agency. Some digital data at 25 000 scale in vector format from the Army Cartographic Services	Municipalities are the main provider of large scale maps. They also co-manage the Cadastre with the Danish Survey and Cadastre and maintain a wide range of other register based information systems

Table 13.1 (cont)

	Greece	France	Poland	Netherlands
Institutional context	Limited financial and manpower resources at the disposal of local government leading to contracting out of IT functions to the private sector and University laboratories. Role of EC funds in facilitating the diffusion of GIS	Long tradition of geographic information handling in large cities. Decentralization measures since 1980 have given new powers to departments and to new administrative structures for main urban areas	Very rapid change from centrally planned to market economy. Earlier attempts to decentralize power to local administration have recently been reversed leaving some uncertainty over the responsibilities of different layers of government	Strong and stable central government with local authorities retaining a high degree of autonomy. Middle layer of provinces struggling to define its role
Structure of local government	Two tiers: 51 prefectures form upper tier, 361 cities and 5600 parishes form lower tier. Average population of parish is 1800	Three tiers: 22 regions, 96 departments and 36 000 communes. Average population per commune is 1500. Variety of intercommunal structures especially in large urban agglomerations	Three tiers: 49 regions, 327 sub-regions, and 2465 local authorities having an average pop. size of some 15 500.	Two tiers: 12 provinces and local government divided between 650 municipalities and 100 polder boards in charge of water control; 50% municipalities have less than 12 000 people and 90% less than 40 000
Digital data availability	Lack of digital data; 5000 scale digital contour data available only for some cities in the Athens and Thessaloniki areas. Proliferation of data produced by private sector data bases	Small scale digital data available from Institut Geographique National. Large scale data in hands of Cadastre. Ongoing digitization programme underway in conjunction with local authorities	Almost half of the county is covered in good quality cadastral information. The process of conversion to digital format has been undertaken but slow progress and organizational difficulties have led a number of local authorities to acquire their own data often in partnership with private sector	Strong awareness of geographic information in the Netherlands and tradition of automated registers (population, etc.). Slow progress of large-scale mapping project based on Cadastre has led to local authorities acquiring thei own digital mapping individually or in partnership with utility companies and private sector

Table 13.2 Key features of GIS diffusion in the five countries where surveys were carried out using a similar methodology.

	Great Britain	Germany	Italy	Portugal	Denmark
Coverage of findings	Comprehensive survey of all local authorities in England, Wales and Scotland in the second half of 1993	Survey of cities with 100 000 population in the first half of 1994	Comprehensive surveys of all regions and provinces in 1993/1994; some additional information for communes (1991) and Metropolitan areas (1993)	Survey of municipalities in late 1993/early 1994	Comprehensive survey of counties and municipalities in first half of 1993
Extent of diffusion	Almost universal in counties and Scottish regions. Half metropolitan districts. One in six shire/Scottish districts	Almost universal: 70/80 cities had GIS, 10 had firm plans	Two thirds regions—one third provinces but another 1 in 3 with firm plans. Limited in medium/large cities	12 municipalities had GIS and a further 24 had AM/FM facilities	Over 80% of all municipalities use register based systems where georeferencing present. GIS/AM/FM almost universal in authorities with 50 000+. GIS in half counties
Geographical spread	North South divide: 32% of Southern authorities had GIS, 24% of Northern authorities had GIS	No significant geographic difference	Pronounced North/South divide among the regions (90% N–37% S) and the provinces (50% N–20% S)	Adoption levels highest in urban areas and in northern parts of Portugal	Urbanized regions generally have higher take up than less urbanized regions
Length of experience	70% systems purchased since 1990	Most systems purchased since 1990	Half regional and 80% provincial systems purchased since 1990	Some municipalities began GIS projects in 1990	Most GIS systems acquired since 1990
Main applications	Automated cartography and mapping for local planning and management	Surveying and topographic data base management	Digital map production, strategic land use planning. Environmental monitoring also strong in the provinces	Land use planning; automated mapping	Digital mapping. Also extensive use for utility management

Table 13.2 (cont)

	Great Britain	Germany	Italy	Portugal	Denmark
Predominant software	Arc/Info in 25–30% of county and metropolitan districts. Axis, Alper Records and G-GP in shire/Scottish/ metropolitan districts	SICAD and ALK-GIAP (latter free to German local authorities)	Arc/Info for over half regional and provincial applications. Greater diversity at municipal level	Intergraph purchased by National Centre for Geographic Information facilities	Intergraph MGE and Autocad based systems with Dangraf and GeoCAD
Perceived benefits	Improved information processing (60%) especially improved data integration, better access to information and increased analytical and display facilities	Improved information processing: (65%) faster information retrieval and increased analytical and display facilities	Improved information processing (51%) especially automated map production and thematic mapping	Improved information processing, administrative reorganization, data sharing with utilities	Improved information processing (66%)
Perceived problems	Technical problems including lack of software and hardware compatibility. Organizational problems especially poor managerial structures and lack of skilled staff	Organizational problems especially lack of qualified staff and insufficient motivation of staff by management	Organizational problems especially lack of awareness, poor coordination and lack of skilled personnel	Bureaucratic inertia, lack of skilled personnel digital data availability, lack of awareness, vendors attitudes and limited follow up support	Technical problems in small municipalities. Organizational problems in large municipalities

Elsewhere in France, Germany, the Netherlands and to some extent Poland, large scale digital data provision is closely linked to the maintenance of the Cadastre. In these countries the establishment of the data infrastructure required for municipal GIS is proceeding rather slowly despite the very strong surveying traditions in some of these countries.

Survey Findings

Table 13.2 summarizes the main findings of the five national surveys. However, before comparing the findings of these surveys, attention must be drawn to a number of differences between them in terms of timing and coverage.

Both the British and Danish surveys involved comprehensive telephone surveys of all local authorities which obtained a 100% response rate. However the Danish survey was carried out in the first half of 1993 between 6 months and a year before the other four surveys. This discrepancy in timing is particularly important given the rapid take up of GIS in the smaller authorities forecast in the Danish chapter of the book.

In Portugal a postal questionnaire was sent to all municipalities in late 1993 but not the regions. It achieved a response rate of 55% which is good for a postal questionnaire but nevertheless the findings must be treated with some caution given that no information is available for non-responding authorities.

Because of the very large number of authorities in Germany, the German survey which took place in the first half of 1994, was restricted to the 86 cities with populations of over 100 000. A response rate of over 90% was achieved in this case which is very high under the circumstances. No information is available for non-metropolitan counties and authorities with less than 100 000 population.

In Italy two comprehensive surveys of the regions and provinces which achieved a 100% response were undertaken in late 1993 and early 1994 respectively. However the information on GIS in municipalities is limited to the findings of research carried out in 1991.

The main findings of the other four studies are summarized in a similar format to those of the five surveys in Table 13.3. It should be noted, however, that these studies are less comprehensive in coverage than the five surveys. In the case of Greece and Poland this presents relatively few problems given the low level of GIS diffusion in these countries. In the case of France and the Netherlands, it should be noted that much of the information is drawn from secondary sources which predate the findings of the five national surveys by several years.

Extent of GIS Diffusion

Tables 13.2 and 13.3 show that the extent of GIS diffusion was lowest in Greece, Portugal and Poland. In these countries between 10 and 15% of the cities had acquired GIS or AM/FM facilities. However, as the case of Portugal indicates, the situation is changing very rapidly as a result of proactive government measures and the extent of GIS diffusion is likely to increase dramatically in the immediate future.

Urban applications in medium and large cities predominate in Denmark, France and Germany. In Germany the use of GIS is almost universal in cities with

Table 13.3 Key features of GIS diffusion in the other four countries

	Greece	France	Poland	Netherlands
Extent of diffusion	About 10% of all cities have GIS facilities	Two thirds of cities with more than 100 000 population have GIS. Widespread use in intercommunal agencies	30–40 mainly in small–medium sized towns (20 000–100 000 inhabitants). Few regional/ sub-regional GIS	GIS is almost universal among municipalities above 50 000 inhabitants and in approx. 50% of those above 20 000
Geographical spread	No marked regional variations	No discernible regional variations reported	No marked regional variations	No marked variations .
Length of experience	Most systems purchased since 1990	Most systems purchased since 1990	Most systems purchased since 1990	Rapid take-up since late 1980s
Main applications	Surveying and topographic data base management	Urban data base management for surveying and planning	Topographic data, parcels data, cadastral information	Mainly topographic and thematic mapping
Predominant software	Arc/Info for over 80% of local government applications	Arc/Info and Apic (French package)	Map/Info, Autocad for small systems, Arc Info for larger central government implementation	Intergraph MGE, Autocad and IGOS (Dutch package)

populations of over 100 000, as it is in Denmark with reference to cities with over 50 000 population. Throughout these countries the extent of GIS diffusion is considerably lower in small towns and rural districts. In France and Denmark it is also lower at the department and county levels.

This relationship is completely reversed in Great Britain and Italy. In Great Britain the take up of GIS in the counties is almost universal whereas only half the metropolitan districts had acquired facilities. In Italy levels of diffusion at the regional and provincial levels are also very much higher than those for the cities. In both cases however the take up in urban areas is generally higher than in rural areas.

In the case of Great Britain the obvious explanation for this difference is the extent to which the surveying and mapping functions that are attached to local government in most European countries are carried out by the Ordnance Survey. In addition there is no requirement in Great Britain to maintain a Cadastre as is the case in most other European countries. The case of Italy, however, is more difficult to explain and may reflect the new planning powers that have been given to the provinces.

Geographical Spread

As noted above a general distinction can be made between urban, especially large urban, and rural areas with respect to the extent of GIS diffusion. In some countries, however, there is also a distinction between regions with respect to the diffusion of GIS. This is most pronounced in Italy with respect to the traditional divisions between the North, Centre and South. At the regional and provincial level GIS adoption has reached nearly 100% and 50% respectively in the North as against less than 40 and 20% for the equivalent authorities in the Centre/South.

There are also clear differences in Great Britain between the North and Scotland and the South and East of the country. In this case, however, the ratio is reversed in favour of the wealthier South and East where the level of GIS adoption is 32% as against 24% in the North and Scotland.

Elsewhere no pronounced regional variations have been reported for Denmark, France, Germany and the Netherlands. In the case of Greece, Poland and Portugal the extent of diffusion so far is probably too low for any clear patterns to have emerged.

Length of Experience

Both Tables 13.2 and 13.3 show that the vast majority of GIS systems in local government in virtually all countries have been purchased since 1990. This highlights the extent to which GIS in local government is very much a recent development in Europe.

Main Applications

Tables 13.2 and 13.3 show that digital map production and digital mapping are the predominant local government applications in Denmark, France, Germany, the

Netherlands and Poland. In most of these countries these activities are closely linked to the maintenance of the Cadastre.

The main exceptions to this rule are Italy, Portugal and to some extent Great Britain. In Italy and Portugal physical planning in one form or another appears to be the main GIS application. This is particularly the case in Portugal where the dissemination of GIS is closely linked to the preparation and approval of municipal land use plans as well as automated mapping. In Great Britain the position is more complex, but the emphasis, nevertheless, is on automated cartography linked to land use planning activities as well as the other technical services provided by local authorities.

Predominant Software

Overall the findings summarized in Tables 13.2 and 13.3 point to the remarkable dominance of North American software, especially Arc/Info, in European local government. Arc/Info itself accounts for over 80% of all GIS applications in Greece, half the regional and provincial applications and Italy and between 25 and 30% of applications in British counties and metropolitan districts. In Portugal, on the other hand, Intergraph is the GIS market leader as a result of its acquisition as part of the implementation of the National Council for Geographic Information's national strategy.

In contrast, the German local authorities have remained largely loyal to their local software developers and the Danish authorities also make extensive use of local software packages. Many of these are custom made for the specific tasks carried out in their local governments. Despite the dominant market share of North American software in most countries there are also a large number of local tailor made packages in use especially at the municipality level. Examples of these include G-GP, Axis and Alper Records (now Sysdeco) in Great Britain, ALK-GIAP in Germany, Apic in France and IGOS in the Netherlands.

Benefits and Problems

Questions regarding the perceived benefits and problems to be derived from GIS were asked only in the five surveys. These revealed a high level of consensus of views throughout all the five countries involved. In the case of benefits there was general agreement that improved information processing was by far the most important benefit to be derived from GIS. The main reasons for this view were improved data integration and faster information retrieval together with increased analytical and display facilities. Irrespective of their nationalities, most respondents also identified similar organizational problems associated with recent developments in GIS. These included poor management structures and bureaucratic inertia together with lack of awareness and the shortage of skilled personnel.

TOWARDS A TYPOLOGY OF GIS DIFFUSION IN EUROPEAN LOCAL GOVERNMENT

In overall terms the findings of the nine studies reveal a considerable measure of agreement regarding the perceived benefits and problems associated with GIS in local

government in Europe. They also show that the length of local government experience
in most cases is very similar between countries. It is useful at this stage, therefore,
to concentrate on identifying some of the main differences between countries and
exploring the extent to which a typology can be developed on the basis of these
differences. From the evaluation above it would appear that two main dimensions
can be identified which account for a large number of the differences observed
between countries.

The first dimension that emerges from the findings is related to the overall extent
of diffusion and the level of digital data availability in these countries. It measures
essentially the links between data infrastructure and diffusion. The second dimension
is associated with the kind of GIS applications that are undertaken and the level of
government at which they are carried out in the different countries. This dimension
largely measures the professional cultures associated with GIS applications.

When the nine countries are plotted in relation to these two dimensions in Figure
13.1 it can be seen that all four quadrants of the diagram are occupied. The largest
grouping consists of countries with high levels of diffusion, reasonable data availability
where GIS is used essentially at the municipal level for surveying and mapping
operations. Denmark is probably the best example of this category, but France,
Germany and the Netherlands possess most of these characteristics.

The second largest group consists of countries with low levels of diffusion and
restricted digital data availability where GIS applications in local government are
predominantly at the municipal level for surveying and mapping. Greece and Poland
are the best examples of this category although Portugal shares some of these
characteristics. However, there is an important difference between Portugal and the
other two countries with respect to the emphasis given to physical planning in the
former. This places it somewhere between Greece and Italy in this table.

Each of the other two quadrants contain only one country. Great Britain stands
out alone as a country with high levels of GIS diffusion and digital data availability
where GIS diffusion is greatest at the upper tier of local government and there is a
strong planning tradition in local government. Similarly, Italy stands out as a country
with relatively poor data and low levels of diffusion where GIS has had its greatest
impact on the regions and provinces also for planning applications.

Two other matters are worth noting from this figure. First, there is a broad
correlation between relative stability and relative change in terms of the institutional

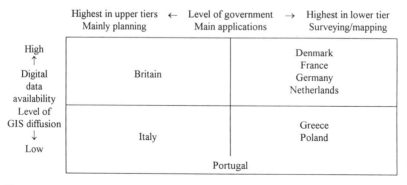

Figure 13.1   Typology of GIS diffusion in nine European countries

context between the upper and lower halves of the diagram respectively. Given the changes that are already in progress in Portugal, for example, and to a lesser extent in Greece, Italy and Poland, the positions of these countries on the table may change considerably in absolute if not in relative terms over the next few years.

Secondly, there are also some interesting links between location in the diagram and the extent of the dependency on imported as against local software. The highest levels of dependency on North American software are generally to be found in the bottom half of the diagram in countries such as Greece and Poland where the overall level of GIS diffusion is still relatively low. In the top half of the diagram on the other hand, levels of dependency are generally much lower and there are also a large number of local software products in use. This is particularly the case in Germany where these products are also the market leaders.

CONCLUSIONS

The findings of this comparative evaluation highlight some of the main similarities and differences between GIS diffusion in local government in the nine European countries whose experiences have been analysed in the preceding sections of this book. They show that in all the countries surveyed GIS are a relatively recent phenomenon in local government and that so far only a limited amount of operational experience has been built up to substantiate the claim that GIS implementation will lead to improved information processing. At this stage it is important to recall the distinction that was made at the outset of the book between adoption, that is the decision to acquire GIS facilities, and implementation, that is the use that is made of these facilities in practice.

GIS Adoption

From the standpoint of GIS adoption it would appear from the analysis that two key factors account for most of differences between the nine countries. The first of these measures the links between digital data availability and GIS diffusion while the second reflects the professional cultures surrounding GIS applications in local government.

The question of digital data availability is not simply a matter of the information rich versus the information poor. It is much more a question of central and local government attitudes towards the management of geographical information. For this reason those countries with relatively low levels of digital data availability tend also to be countries where there has been a fragmentation of data sources in the absence of central government co-ordination. Conversely those countries with relatively high levels of digital data availability tend also to be countries where governments have created a framework in terms of responsibilities, resources, and standards for the collection and management of geographic information.

The question of professional cultures is closely linked to the responsibilities that have been given to local government in different European countries. In countries where local governments play an important role in maintaining land registers and/or cadastral systems, a highly organized surveying profession has come into being to carry out these tasks. Consequently, land information systems rather than geographic information systems tend to predominate in local government applications in these countries. Conversely, in countries such as Great Britain where land registration and digital topographic mapping are dealt with centrally, there is a greater emphasis on applications which are broadly linked to the various planning activities of local government. In this case the predominant professional cultures tend to be those of the land use planner and, to a lesser extent, the transport engineer.

Within these two groups, however, it is also necessary to take account of the nature of the responsibilities that have been given to local government and the resources placed at their disposal when comparing professional cultures. There are considerable differences, for example, as Craglia (1993) has shown elsewhere between local planning functions and associated cultures that have developed in Great Britain and Italy. Similarly there are also important differences between Germany and Greece with respect to the resources at the disposal of local government for land management functions which affect the development of the surveying profession in each country.

As a result of this analysis it can be argued that these two factors provide a useful starting point for further research on the diffusion of GIS in local government in both Europe and elsewhere. What is now needed is further studies which evaluate and refine the typology developed for the analysis in the context of local government in other countries. There is also a need for cross national comparative studies of other key GIS application sectors. In this respect special priority should be given to the utilities sector because of its central position as the other key investor, alongside local government, in GIS technology.

The final question concerns the significance of the findings of the analysis of GIS adoption with respect to diffusion theory. On this count it shows that there are considerable differences between countries with respect to their positions on the S shaped curve that was described in Chapter 1. GIS diffusion in local government in Great Britain, Denmark and Germany in particular is already well past the critical point of take off and is approaching saturation in relation to particular levels of government. Elsewhere, in Greece, Poland and Portugal in particular, GIS diffusion in local government is still at a relatively early stage on the curve, but, as the findings of the survey clearly indicate, the situation is changing very rapidly at the present time. Given the speed of change it will be necessary therefore to closely monitor events to keep track of the adoption process in these countries.

GIS Implementation

It must be recognized that the primary focus of the above discussion has been on GIS adoption and that matters relating to GIS implementation and its consequences have only been discussed in general terms in Part I of this book and in some of the case studies in Part III. Nevertheless it is clear from this discussion that implementation is likely to become an area of growing significance in GIS diffusion research.

GIS implementation research is primarily concerned with the processes whereby new technologies are adapted to meet the specific needs of organizations such as local authorities. The starting point for such research, as Heather Campbell's chapter demonstrates, is not classical diffusion theory but theories of organizational change. Given that the focus of this research is on the organization itself rather than on the national circumstances within which that organization operates, it may be expected that some striking similarities will be found between organizations in widely differing institutional contexts. In this way the findings of such research are likely to further highlight the diversity of responses to the adoption and implementation of new technologies such as GIS.

However, implementation research may also have a wider scope. As the contribution by Michael Wegener and Ian Masser indicates, the widespread introduction of GIS or GIS functions in manufacturing, services, retailing and leisure activities, logistics and travel, government and public and private planning may have consequences far beyond their direct impacts within their narrow field of application. GIS are already now indispensable for maintaining the efficiency of modern economies and the convenience of modern life and are expected to continue to open up new opportunities of economic and social activity. But there are also fears that GIS may be used to undermine essential civil rights and societal values, create new patterns of dominance and dependency or impose on their users a certain view of the world which may be alien to their culture. These wider social and political consequences need to be investigated in more detailed social-science based studies. To develop a conceptual framework for such studies is the objective of a future activity in the GISDATA programme focusing on the role geographic information in the Information Society.

ACKNOWLEDGEMENT

The authors of this chapter would like to express their appreciation to the nine case study contributors for their comments on an earlier draft of this chapter and for the additional information that they provided to facilitate the comparative evaluation. Nevertheless, the responsibility for interpretation of the findings remains with the present authors.

REFERENCES

CRAGLIA, M., 1993. Geographic information systems in Italian Municipalities: a comparative analysis, Doctoral dissertation, University of Sheffield: Department of Town and Regional Planning.

Index